THE COLD WARS

THE
COLD WARS

A History of Superconductivity

Jean Matricon and Georges Waysand
Translated by Charles Glashausser

Rutgers University Press
New Brunswick, New Jersey

Originally published as *La guerre du froid: Une histoire de la supraconductivité*,
© 1994 Editions du Seuil

Library of Congress Cataloging-in Publication Data

Matricon, Jean.
   [Guerre du froid. English]
   The cold wars : a history of superconductivity by Jean Matricon and Georges Waysand ; translated by Charles Glashausser.
      p.   cm.
   Includes bibliographical references and index.
   ISBN 0-8135-3294-9 (hardcover : alk. paper) — ISBN 0-8135-3295-7 (pbk. : alk. paper)
   1. Superconductivity—History.   I. Waysand, G.   II. Title.
   QC611.95.M3813   2003
   507.6′23′09—dc21

2002152303

British Cataloging-in-Publication information is available from the British Library.

The publication program of Rutgers University Press is supported by the Board of Governors of Rutgers, The State University of New Jersey.

Cet ouvrage, publié dans le cadre d'un programme d'aide à la publication, bénéficie du soutien du Ministère des Affaires étrangères et du Service Culturel de l'Ambassade de France aux Etats-Unis.

This work, published as part of a program of aid for publication, received support from the French Ministry of Foreign Affairs and the Cultural Services of the French Embassy in the United States.

Manufactured in the United States of America

CONTENTS

PREFACE

Prehistoric man learned how to make heat when he discovered how to make a fire, and he left to his wife the serious task of maintaining it. At the beginning of the nineteenth century, however, no one knew how to produce cold. Ice was familiar, of course, and useful, but there's a big difference between making use of a natural phenomenon and making the phenomenon happen. We use rainwater for irrigation, but we can't make it rain at will.

Progress in thermodynamics, the science of heat, early on revealed a basic asymmetry between heat and cold. It is physically possible to reach temperatures of many thousands of degrees, but it is meaningless to speak of temperatures of minus thousands of degrees. At −273.16°C, far below zero on standard thermometers, there's a lower limit on temperature called absolute zero. There is no temperature below absolute zero, and, in fact, absolute zero itself is impossible to reach. Getting as close as possible to absolute zero has been an intellectual quest for several centuries; scientists always expect, or, at least, hope to find new phenomena as they explore new territory. In addition, the search for low temperatures provides new ways to study the structure of the matter that surrounds us, just as important as those cathedrals of contemporary physics, the big particle accelerators.

Devices capable of reaching temperatures close to absolute zero are complex, but they are small in size compared with accelerators and fusion reactors. These huge devices reproduce physical conditions that have already existed somewhere, extreme yet relatively simple conditions like those that existed just after the big bang or that still exist at the core of stars. The approach to absolute zero, on the other hand, with rather more modest means, has created something that probably never existed before the twentieth century in the entire universe: temperatures lower than the temperature of the primordial radiation from the big bang that fills our entire universe and bathes our planet. This ubiquitous radiation ensures that no isolated object can maintain itself at a temperature less than 2.735 centigrade degrees above absolute zero (called 2.735K, where K stands for kelvins); the object will heat up inexorably until it reaches this equilibrium temperature.

This new world of low temperatures, in which the fog of thermal agitation has lifted, is not immobile or even necessarily frozen. Little by little, low-temperature physicists have discovered many types of motion ordered in surprising ways. While particle physicists search for the fundamental constituents of matter, mostly with accelerators, they tell us little about how matter is organized. Yet the properties of the whole are seldom the sum of the properties of the parts. The physics that opens up the world of low temperatures is the physics of highly organized systems and their transitions from one type of organization to another.

This book is devoted to *superconductivity*, one of the phenomena encountered at low temperatures that has fascinated physicists ever since its discovery in 1911. It has turned out to provide a fruitful arena for development of the subtle theories and methods of the general field of condensed matter physics, which have proved applicable in particle physics and cosmology as well. Until recently, superconductivity has attracted a limited audience, in spite of the Nobel Prizes awarded for each major advance. The situation changed completely with the discovery in 1986 of materials that were superconducting at temperatures well above absolute zero, contrary to previous expectations. The history of superconductivity is thus far from being closed, but it is no longer necessarily the realm of temperatures close to absolute zero.

We have presumed that readers are as curious about the history of ideas as they are about the nature of matter. In fact, superconductivity provides a classic example of the evolution of physics in the twentieth century: the science itself, its epistemological foundations, and its social context. This book should be accessible to readers with only a basic background in science; it is superconductivity for the layman, a description of the undeniable successes of the field. Beyond that, however, we want to achieve some reasonable perspective about the development of the field and its relationship with the rest of physics and the history of our time. Our colleagues in physics can read it with a red pencil in hand to keep track of our errors, simplifications, and prejudices, if they wish, provided they also question their own assumptions about the goals and conditions of our discipline. In so doing, they will only be following the eminent English physicist James Clerk Maxwell, who told us that "the history of science must tell the story of investigations that have not succeeded. . . . The history of the development of ideas be they normal or abnormal is of all subjects the one that we, as humans, are interested in the most."[1]

Superconductivity was such a surprising phenomenon that it drew the attention of many talented physicists, who were successful in explaining it only after a half-century. This delay allowed superconductivity to confirm its own uniqueness within solid-state physics (a branch of physics that itself did not become autonomous until the 1930s and has been known more recently as condensed matter physics). Even though the field of superconductivity has existed longer than quantum mechanics, critics have paid less attention to it. Thus superconductivity provides fresh territory for the examination of the internal evolution of a discipline, the connection between theory and experiment, the relationships among the various subfields of physics, and, finally, the interactions and (often unintended) quarrels between science and society.

Superconductivity had little influence on society but was strongly influenced by political events of the last century. The scientific, epistemological, and social aspects of superconductivity are reflected in this book. Our perspective is not linear; this is not a teleological view of the evolution of the field. Our voyage will take us across Europe and thence to America and all the industrialized countries, so it may be useful here to indicate the principal stops on this trip.

The idea of absolute zero is older than one might think, but experimenters did not get close to achieving it until the second half of the nineteenth century, which is where we begin our story. In a climate of growing competition, gases were being liquefied one after the other—first oxygen (in France and Switzerland), then hydrogen (in England and Poland). Helium was the last, and this development required a conservative with an iron fist who brought about a radical change in the style of scientific work. Heike Kamerlingh Onnes applied industrial methods to science: he spent a quarter of a century constructing in Leiden, the Netherlands, the predecessor of all big science laboratories. Helium was liquefied there for the first time in 1908, and three years later superconductivity was discovered—the electrical resistance of mercury fell to zero at 4.2K. Explaining superconductivity required another half-century. The theory of electricity and heat transport available in 1911 was in good accord with experiment, but its simplicity made it suspect. Only the appearance of quantum mechanics in 1925 led to something better.

The quantum theory of metals had some immediate successes, but superconductivity resisted the best efforts of a superb group of prestigious physicists, even suggesting a theorem: "Every theory of superconductivity is a false theory." To understand the nature of superconductivity, the properties of a superconductor had to be known better—was it anything more than a perfect conductor? A response was delayed by experimental problems in determining both the magnetic properties and the thermal properties of superconductors. When the Leiden monopoly on liquid helium was broken, W. Meissner in Berlin showed that the magnetic field in the interior of a superconducting metal was rigorously zero, a property that demonstrated that a superconductor was radically different from a perfect conductor. Two Dutch physicists, H. Casimir and C. Gorter, were the first to understand this distinction, which they translated into a thermodynamic model. The theoretical spectrum was broadening, but Europe was beginning its turbulent years.

Fritz London, a German theoretical physicist, almost simultaneously provided the first model of the superconducting state that had predictive power, particularly for Meissner's results. It was based on macroscopic quantum order and yielded a simple understanding of the electrodynamics of superconductors.

The liquefaction of helium lowered the temperature limits for the exploration of matter, including helium itself, where a phenomenon akin to superconductivity was discovered at 2.19K in 1938 — in Cambridge in spite of Stalin and in Moscow because of Stalin. The Soviet Union had begun scientific activities again in the 1930s and had chosen low-temperature physics as one of its priority fields. An entire generation of Soviet physicists, among them such distinguished figures as Lev Shubnikov, Lev Landau, and Piotr Kapitza, obtained remarkable results under tragic conditions, certain aspects of which were revealed only in 1993.

The rise of the Nazis had provoked a mass exodus of low-temperature physicists that started before the war. The outbreak of worldwide conflict and the role that physics was called upon to play in this conflict had numerous unexpected consequences even in the limited domain of low-temperature physics. The techniques of super high frequencies that came out of radar and the access to isotopes furnished by nuclear reactors just after the war revealed two essential traits of superconductivity: the nonlocality of phenomena related to electrons and the important role of the ions in the crystal lattice.

A situation without precedent in the history of physics arose in this context: the cold war entirely stopped the flow of scientific communication between the Soviet Union and the other industrialized countries. In the East, Vitali Ginzburg and Landau brought to fruition a phenomenological theory that proved to be a remarkably efficient tool. This result was almost totally unknown in the West, where several groups of theorists were trying to determine the mechanism responsible for the appearance of the superconducting state. In 1957 John Bardeen, Leon Cooper, and J. Robert Schrieffer found the key: electron pairing via coupling to the quantized vibrations of the crystal lattice. London's intuition about a macroscopic quantum mechanical order was justified.

It was the golden age of superconductivity. The wounds of World War II were healed; scientific expansion in tune with economic expansion seemed limitless. With the East-West détente, scientific exchanges took up a more normal rhythm. A bountiful harvest of scientific results and ideas for applications of superconductivity followed, but in the 1970s an economic crisis, research and development programs that were a bit too baroque, and difficulties with cooling apparatus tended to slow the pace of progress. Even improved microscopic theories offered little aid to experimentalists trying to raise the temperature at which superconductivity appeared. The researchers were numerous, but guiding principles were rare.

Superconductivity became one more branch of materials research, and the physicists were in competition with inventive chemists. In a climate of increasing gloom, which was afflicting most branches of physics, the announcement of superconducting ceramics at 35K and soon thereafter 100K started a revolution. The unprecedented publicity surrounding these events led one to hope for miracles. In the middle of the 1990s these miracles have not yet occurred, but the concept of a room-temperature superconductor no longer appears to be in the realm of science fiction, even if the mechanism that might produce it is unknown.

After almost a century the breadth of knowledge about superconductivity largely surpasses the limits of the original field. Now that the chase for patents so often stifles efforts at comprehension, it is unlikely that the field will ever be the same again, although it is likely to offer other surprises.

We very much appreciate the support of numerous colleagues who agreed to be interviewed or took the time to furnish us with documents that were difficult to trace. In the first rank we thank Jean-Paul Mathieu, now deceased, who throughout his life defended his physics colleagues who were under attack for their political views; he provided us with precious elements of documentation. We are grateful also to Nicolas Kurti of Oxford, a member of the Royal Society, for communicating the results of his investigation of the liquefaction of oxygen and for his information concerning cryogenics at very low temperatures and the introduction of helium liquefaction to France, in which he participated. We thank Joseph Lajzerowicz for his information on the posthumous publication of the works of Lavoisier; Hendrik Brugt Casimir, a member of the Netherlands Academy of Sciences and past director of the Philips Laboratories, to whom we are indebted for much of our information about Leiden, from his own book of recollections and from an interview which he kindly granted; and R. De Bruyn Ouboter, who, in addition to his own scientific activity, devotes time to analyzing the archives of the Kamerlingh Onnes Laboratory and who kindly communicated the fruits of his work and discussed with us the early years in Leiden.

We thank Françoise Balibar for providing us with the correspondence between Einstein and Ehrenfest; L. F. Feiner (Philips Laboratory, Eindhoven) and P. M. Claasen, a grandson of Willem Hendrik Keesom, for providing the seminar notes of his grandfather; Martin Ruhemann for his recollections of the Kharkov laboratory; David Shoenberg of Cambridge for his comments on his relations with Kapitza and for providing copies of Kapitza's letters before publication; David Rousset and Théo Bernard for documents concerning Alexander Weissberg-Cybulski; to Professor Krasowski of the Ukrainian Academy of Sciences for furnishing a copy of the posthumous book of homage to Shubnikov; Igor Dzyaloshinskii, a colleague of Lev Landau, for extracts of Landau's KGB dossier; to Rudolf Mössbauer and Klaus Andres, director of the Meissner Institute in Munich, for information about their institution's namesake; to Jacques Friedel, of the French Academy of Sciences, for his remarks on the French scientific context; to Mrs. Edith London and Horst Meyer for providing a photo of the London brothers; to Vitali Ginzburg for kindly granting us an interview about his work with Landau; to Alexeï Abrikosov for his personal recollections; to Robert Schrieffer for his unpublished comments about the Seattle conference; to Luke Chu Liu Yuan, who, in the course of a long scientific collaboration with one of us and numerous dinners shared in good humor, has provided information and documents on the contribution of Chinese scientists to American scientific life; to Sophie Benech, Wladimir Berelovitch, and Daniel Singer for their information on political life in the USSR; to Jean-Louis Sabrié (Alsthom) and Elie Track (Hypres) for information on several technical applications of superconductivity; to J.-G. Le Gilchrist and P. Monceau for their aid in our iconographic research; finally, to a great many other colleagues who have drawn our attention to various aspects of this long history, particularly Christiane Caroli and Daniel Saint-James with whom we discussed the golden age of superconductivity on several occasions.

Throughout the preparation of this book, we benefited from the facilities of our laboratory, the Groupe de Physique des Solides of the Université Denis-Diderot (Paris VII) associated with the Centre National de la Recherche Scientifique (CNRS), which we thank. Finally, the authors are indebted to their editor and friend Jean-Marc Lévy-Leblond for having waited five years for this manuscript and for having returned it "quantumly correct," as well as to our attentive readers, Christiane Caroli and Sylvain Hudlet; to Isabelle Wagner and Veronique Ovaldé, who knew how to put an end to this long wait by managing with competence and kindness the fabrication of this book. In spite of this, of course, all remaining errors of fact or judgment are solely our own responsibility.

—Jean Matricon and
Georges Waysand, Paris,
August 1994

TRANSLATOR'S NOTE

Having read and enjoyed *La Guerre du froid* at the behest of Helen Hsu, then science editor at the Rutgers University Press, I recommended publication and suggested my colleague, Peter Lindenfeld, an expert in superconductivity and a friend of the authors, as translator. His recent retirement, however, had only expanded the number and range of his activities, leaving no time for such a project. I then offered to do it, my genuine enthusiasm limited only by my unfamiliarity with the field and my lack of experience in translation. Peter agreed, however, to comment on my drafts (remarkably astutely in the event) and to advise me on the sometimes arcane physics, language, and methods of superconductivity research. Peter was, as always, an impish delight to work with; he provided indispensable help. He encouraged an economy of words and a straightforward approach to the explanation of ideas that led to occasional deletions, additions, and rearrangement of parts of the original text.

Jean Matricon and Georges Waysand have been my close collaborators. We have spent weekends together in the sunnier parts of France as well as grayer days in the Tour Jussieu in Paris; they have listened to my reading of the entire manuscript, laptop and *La Guerre* at hand, generously accepting (or insistently rejecting) my interpretations. My linguist friend Anthony Arlotto made numerous suggestions after spending several weeks with the manuscript in the summer of 2002; Audra Wolfe commented in detail on the first chapters after her arrival as science editor at the Press in 2002. My colleague Harry Kojima explained some particularly delicate aspects of superconductors in magnetic fields. To all I am deeply grateful; in spite of their guidance, I have certainly persisted in errors in translation for which I alone am responsible. Like Douglas R. Hofstader, the author of the charming exploration of the meaning of translation, *Le Ton beau de Marot*, who unknowingly provided practical advice and inspiration for my perseverance in this work, I suffered a grievous loss midway through. This translation is dedicated to the memory of my wife, Suellen.

"Serious work is sterile."
—*Georges Bataille*

THE COLD WARS

CHAPTER 1

The Logic of Low Temperature

A trace of the wonder surrounding the discovery of superconductivity persists even today. Books about it begin like fairy tales: "In Leiden in 1911, Kamerlingh Onnes..." Only the "Once upon a time" is missing. The discovery that the resistance of mercury suddenly drops to zero around −269°C, close to what is now called absolute zero, was not a trivial accomplishment. The first milestone was reaching this temperature, the temperature of liquid helium at atmospheric pressure. Even in the most rigorous Siberian winter the temperature only very rarely gets below −60°C. The history of superconductivity begins in Leiden because it was the only place in the world at that time where anyone knew how to create such intense cold. The ritual was carried out every week, driven only by scientific curiosity.

The theoretical basis for an absolute temperature scale is commonly attributed to Lord Kelvin. Most physicists believe that credit for the idea of an absolute zero of temperature belongs to J. A. C. Charles and J. L. Gay-Lussac, who showed by a simple extrapolation of their measurements that the volume of any gas would shrink to zero near −273°C if the pressure was held constant. Even earlier, however, Guillaume Amontons, who was born in Paris in 1663, seems to have been the first to come up with this concept, well before the birth of thermodynamics in the nineteenth century.[1] Amontons noted that the pressure of a constant volume of air decreased by the same amount for each degree that he lowered the temperature. As a result, he estimated that the pressure would go to zero near −240°C. This zero was purely hypothetical until around 1850. Step by step, however, scientists would get closer and closer in the years that followed.

Establishing a temperature scale requires a choice of units. In most countries, except the United States, the standard unit is the centigrade degree, obtained by dividing the temperature interval between the melting point of ice and the boiling point of water at atmospheric pressure into one hundred equal parts. Both the Celsius scale and the Kelvin scale use centigrade degrees. On the Celsius scale the temperature of the ice is labeled 0°C and the temperature of the steam 100°C. One hundred centigrade degrees on the Kelvin scale still separate the temperatures of the melting point of ice and the boiling point of water at atmospheric pressure. But the zero on the Kelvin scale is absolute zero, which happens to be 273.16 centigrade degrees below the temperature of the melting ice. Thus, the temperature of ice on the Kelvin scale is 273.16K (and steam is 373.16K). We can convert from one scale to the other with the equation T (kelvins, K) = T (degrees Celsius, °C) + 273.16. The conversion between the Fahrenheit scale used in the United States and these scales is slightly more difficult because a Fahrenheit degree is smaller than a centigrade degree, and the zero has no direct physical significance. Thus, absolute zero is −459.7°F, ice melts at 32°F, and water boils at 212°F.

The success in liquefying helium in Leiden in 1908 was the last step in a vast nineteenth-century enterprise, the liquefaction of gases. What started off as an intellectual exercise became one of the century's great driving forces of technology. Louis Cailletet's machines and James Dewar's containers, examples of the many instruments and special devices that had to be developed, were crucial to the discoveries described later in this chapter. The achievements were not merely technical, however; essential theoretical advances accompanied each step.

Michael Faraday (1791–1867) was the first scientist to become interested in liquefying gases. His fame is tied more closely to his work in electromagnetism (the *farad*, e.g., is a unit of capacitance). In 1823, however, he succeeded in compressing gaseous chlorine until it liquefied in the closed end of a V-shaped test tube. Simply compressing any gas, it seemed, should yield a liquid. In fact, Faraday liquefied "almost all the gases" (and discovered benzene in oil tar), according to *Le Petit Larousse*. What is not said in this excellent book is that Faraday realized very quickly that his failure to liquefy all gases was not an accident. He turned this failure into a success when he coined the term *permanent gases* to label his discovery of gases that could not be liquefied by compression. The designation "permanent gases" became quite widespread; among them were oxygen and hydrogen, both essential for chemistry and industry. Compressed to three thousand times atmospheric pressure, they remained gaseous even as they got more and more dense.

Economic interests were now at stake in this activity, even though scientists like Faraday did not raise such issues. In 1856 Henry Bessemer had suggested using oxygen to improve steel refining. Large quantities of the pure gas would be needed, and liquefaction was known to be a way to purify it. Before the first drop of liquid oxygen was made, then, the process already had industrial applications. Such interplay between science and industry will often capture our attention in this book. In fact, steelmaking with pure oxygen did not become an industrial reality until well after World War II, when it indeed yielded better steel, without carbon contaminants. Bessemer's suggestion is a well-known example of an important technical idea that arrived before

Compression and liquefaction of chlorine by Faraday. Because its high critical temperature (144°C) allows liquefaction at low pressure and moderate temperature, chlorine was the first gas to be liquefied. Faraday achieved this result in 1823 by enclosing compressed chlorine gas in a V-shaped tube and immersing one side in a coolant.

its time. His idea was carried out much later than one might have thought, primarily because of the large amount of capital that is needed for steelmaking.

Physicists in the middle of the nineteenth century understood that it wasn't just the pressure that was important in liquefying gases; it was also the temperature. They thought that, if they could lower the temperature sufficiently, they might liquefy oxygen. In 1852 James P. Joule and William Thomson (Lord Kelvin) each showed independently that a sharp decrease in pressure led to rapid cooling, but the pressures they could achieve were not sufficient for conclusive tests.

In 1863 Thomas Andrews, an Irishman, following up on the work of Charles Cagniard de La Tour (the inventor of the siren), provided a detailed description of the necessary conditions for gas and liquid states to coexist. Cagniard, an attaché in the Ministry of the Interior in Paris (the ministry responsible for the police, thus the siren), had suggested that there exists a temperature above which a gas can never be liquefied, no matter how high the pressure. This temperature is called the *critical temperature* (for liquefaction); above it, the liquid phase cannot exist, even as tiny droplets. Andrews showed experimentally that this was true for carbon dioxide, and he surmised that it was true for all gases.

The problem that had already grabbed Andrews's attention, one that would capture more and more physicists interested in condensed matter, was understanding phase transitions, the passage from one phase to another. The term *phase* designates a state of matter, here liquid or gaseous, later on resistive or superconducting. The physical quantities that define the state of a gas are its temperature and pressure. For a given amount of gas at a fixed temperature, there is a relation between its pressure

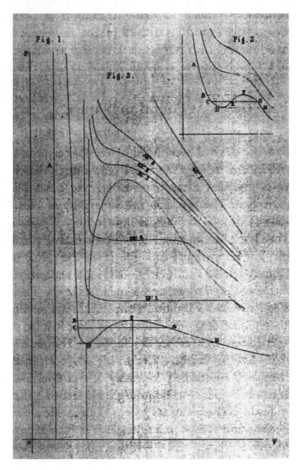

Andrews's isotherms. These are curves that show the relation between the pressure P in a gas versus its volume V when the temperature T of the gas is maintained constant. For a so-called ideal gas, these curves are hyperbolas, determined from the ideal gas law, $P = RT/V$, where R is a number called the ideal gas constant. Examples can be seen in the upper middle of the figure. The curves shown here are for an actual gas, CO_2, for which the situation is more complicated than for an ideal gas, primarily because it can be liquefied. At temperatures below the critical temperature for liquefaction, the isotherms include a zone in which both the pressure and the temperature do not vary as the volume changes; this is the region in which the liquid and the gas phases coexist. The isotherms are then flat, horizontal lines as seen in the bottom of the figure. Mapping out such isotherms for all gases, including those that were the most difficult to liquefy, like hydrogen and helium, was an important activity for physicists at the end of the nineteenth century, particularly in the Netherlands.

and its volume that can be represented by a curve on a graph of pressure versus volume. The curve is called an *isotherm* (equal temperature) because the temperature is the same everywhere along it. Andrews plotted many isotherms for carbon dioxide at different temperatures. He noted that they looked very much alike below the critical temperature but very different from those above it.

For all temperatures below the critical temperature, simple compression, generally to pressures less than a hundred times atmospheric pressure, will liquefy a typical gas. From the moment some liquid begins to form, the pressure remains constant as long as some unliquefied gas remains, and the volume decreases as the material becomes more dense. If we plot this (constant) pressure against this (decreasing) volume, we get a horizontal straight line. The change in volume is larger at lower temperatures, so that the length of the straight line on the plot is then longer. At higher temperatures, however, the line gets shorter and reduces to a point as the critical temperature is reached. We are all familiar with the idea that a physical quantity might stay constant during a phase transition. It happens during a change in phase that we know well, the liquid-to-gas transition that occurs when we boil water for coffee in the morning. As long as some liquid remains, the water temperature stays constant while it boils, no matter how high we turn up the heat.

Discovering that a critical temperature for liquefaction exists did not solve the problem of liquefying oxygen, which remained a complete mystery. At best one could imagine that it would require a clever mixture of temperature and pressure. Liquefying oxygen was becoming a scientific challenge, and passions were rising.

Popular writers paid attention. In1874 Jules Verne, one of the most famous of the century, published a surprising novel, *Doctor Ox*. The good doctor is called upon by the citizens of the city of Quiquendone to pursue his work in their pleasant town, lazily ensconced in a valley with a tranquil stream. Quiquendone bears a striking resemblance to the actual town of Amiens, a small town in the French province of Picardy, where Jules Verne lived a large part of his life. Our doctor has a hard-working assistant, Ygene, who helps him set up a major piece of apparatus full of tubes. The Quiquendonians, beer drinkers and pipe smokers all, had no idea what the pair was up to. Little by little, they lost their legendary cool, and irritability became their daily lot. Lovers, friends, everyone quarreled with everyone else. No one knew why. Then one day two of the townspeople climbed up to the top of the bell tower. Once there they forgot why they had come, so surprised were they to discover that their good humor had returned. And why, indeed, did they feel good again? Perhaps the reader has already guessed. What Ygene had constructed, based on the plans of Ox, was something that looked like a brewery, but, instead of making beer, it made an immense invisible cloud, of Ox-Ygene! Oxygen, denser than air, had spread out over the entire valley and stalled there. Only by climbing up the bell tower could the Quiquendonians breathe normal air once again.

For once reality didn't take long to catch up with fiction. At the end of 1877 oxygen was finally liquefied. The circumstances of the event were, when all was said and done, even more farcical than the misadventures of the Quiquendonians. The site was the slopes of the "Mountain" of St. Genevieve, the section in Paris where both the Ecole Polytechnique and the Ecole Normale Supérieure were located and which the authors of this book know well.

On December 24,1877, invigorated by a sense of duty few institutions these days possess on Christmas Eve, the members of the Academy of Science gathered under the chairmanship of Jean-Baptiste Dumas, their permanent secretary. Dumas, a celebrated chemist, is credited with having determined a large number of atomic weights as well as a monograph, *Treatise on Chemistry Applied to the Arts,* which *Larousse* (the French *Merriam-Webster*) calls "one of the most beautiful monuments of the science of chemistry." A large part of the meeting was devoted to papers on phylloxera, a disease attacking the vineyards and thus a critical issue at the time. But Dumas also read a note from Cailletet, who had just been elected, one week earlier, as a corresponding member of the academy. The subject of his communication was the liquefaction of oxygen, and we read in the *Proceedings of the Academy of Science:*

> Chemistry. Concerning the condensation of carbon dioxide.
> NOTE FROM M. LOUIS CAILLETET.
> If one encloses some pure oxygen or carbon dioxide in a tube of the shape that I have described and puts it into the compression apparatus that I have demonstrated for the Academy; if one then cools this gas down to a temperature of $-29°C$ with sulfurous acid and raises the pressure to about 300 atmospheres, these two gases remain in the gaseous state. But if the pressure is then suddenly relieved, which, according to Poisson's formula, should make the temperature drop at least 200°, immediately an intense mist appears, produced by the liquefaction and perhaps the solidification of the oxygen or the carbon dioxide.

This was but the first dramatic moment of the meeting. To general amazement Dumas then announced that a communication had been received from M. de Loynes, the representative of the Paris branch of Raoul Pictet and Co. The heading on the letter from M. de Loynes noted that the company made "apparatus for producing low temperatures and ice starting from anhydrous sulfurous acid." According to the *Proceedings of the Academy of Sciences:*

> Chemistry. Experiments by M. Raoul Pictet on the liquefaction of oxygen, communicated by M. de Loynes.

> *We have the honor of reporting to the Academy a communication concerning an important result recently obtained by M. Raoul Pictet in Geneva.*
> *On December 22 at 8 P.M. in the evening, we received from him the following dispatch:*
> *Oxygen liquefied today at 320 atmospheres and 140 degrees of cold by combining sulfurous acid and carbon dioxide.*

> SIGNED: RAOUL PICTET

We have since received, in addition, several explanations that we attach here concerning the procedure employed by M. Raoul Pictet to obtain this result which he had been trying to achieve for a long time.

In other words, as often happens when a scientific program has become established as a paradigm, two independent investigators reached a well-defined goal simultaneously. Not unexpectedly, there was a quarrel about who was first. At the academy that day, just as at Quiquendone, there was excess oxygen in the air. Henri Sainte-Claire Deville, another illustrious chemist, stood up; one can only wonder whether this scenario for the meeting had been planned in advance:

> M. Cailletet repeated his experiments on the condensation of oxygen at the laboratory of the Ecole Normale on Sunday, December 16. The experiments were a complete success, in perfect accord with the description we just heard in his communication. His note was not published sooner because M. Cailletet was a candidate for the position of corresponding member that the Academy awarded him during the meeting on December 17. We discussed his qualifications on December 10. He did not want to bring forward at that moment a work whose results had not yet been confirmed by an experiment witnessed by competent judges. Finally, it didn't seem appropriate to him to publish something admittedly very important on December 17, the day of his election, when it had not been discussed in the secret committee on December 10. Fortunately, however, on December 3, I had taken the precaution of having the permanent secretary sign and seal a letter containing both an announcement of the discovery and a confidential statement about his honorable feelings at this time. Thus priority belongs incontestably to Cailletet. But I have to add that the remarkable work of M. Raoul Pictet is not at all tainted by this. The experimental method he followed is totally different from the procedure used by M. Cailletet.

Dumas then read the letter from Cailletet to Sainte-Claire Deville dated December 2, which the latter had sealed on December 3:

> I must tell you, you first and without losing a single instant, that this very day I have liquefied carbon dioxide and oxygen.
>   Perhaps I am wrong to say liquefy, because at the temperature I can reach by evaporating sulfurous acid, $-29°$ and 300 atmospheres, I do not see a liquid but rather a mist thick enough for me to conclude that a vapor very near condensation must be present. I am writing today to M. Deleuil to ask him for some nitrogen proto-oxide. With that I will surely be able to see a flow of carbon dioxide and of oxygen.

> P.S. Just a few minutes ago I did an experiment which puts me quite at ease. I compressed some hydrogen to 300 atmospheres, and, after cooling it to $-28°$, I let it expand quickly. There was not a trace of mist in the tube. My gases (CO and O) are thus close to the condensation point.

Nicolas Kurti, a famous cryogenicist of our own day, was a sometime historian of the sciences and very familiar with academic customs. Not satisfied with the official version of the proceedings, he went back to the original documents and found that matters proceeded in a manner slightly less idyllic. First of all, the permanent secretary had (intentionally, we think) censored the communication from M. de Loynes, Pictet's representative, concerning Pictet. Thinking he was doing the right thing, de Loynes had pointed out in his conclusion what Pictet, a Swiss citizen, owed to France: "France can be proud that M. Raoul Pictet took his degrees at the Ecole Polytechnique and at the College de France. Surely you are aware that he works in industry, making anhydrous sulphurous acid. This is the very product that has made it possible for him to overcome his previous difficulty and get down to low enough temperatures."

By the same token, more personal passages were expurgated from the letter that Cailletet sent to Sainte-Claire Deville. This is what Cailletet actually said:

> Here is what I would like to do. If my election to the Academy takes place soon, *i.e.*, before the end of December, I'll go to Paris and present the complete results to the Academy at that time. If, however, the election will not take place until later, I would be obliged if you would let me know. Then I'll go to attend the meeting next Monday if you think it useful. In any case, I would appreciate it if you would not speak about these results that I am so happy to tell you about and that could perhaps be useful for my candidacy. Talking about them would let the cat out of the bag.

In other words, Cailletet had wanted exactly the opposite of what Sainte-Claire Deville claimed. Finally, Kurti also found the letter in which Sainte-Claire Deville requested of Dumas that the sealed letter be opened during the December 24 meeting, since the results of Pictet were already known. Sainte-Claire Deville's letter reveals his real motivation:

> A scientist very much a Polytechnician believed he should take the bloom off the rose of Cailletet's discovery. This would be harmful to the Academy, to which this fellow doesn't belong, and to Cailletet, who has ties with the Ecole Normale. This little fact came out yesterday at the meeting of the society [unreadable]. The newspapers are full of it. Please inform the Academy about Cailletet's note or read it aloud. And that letter from Cailletet which you signed and sealed — please have it opened in front of the Academy. Even though I am half Genevan, I hope that I can help you in giving our compatriot that which legitimately belongs to him without in any way detracting from the merit of M. Pictet. I strongly insist on telling you what happened last Sunday at the Ecole Normale, and on explaining the honorable reasons which kept Cailletet's mouth shut for almost a month about such an important subject.

L'Ecole Normale versus L'Ecole Polytechnique — the south face versus the north face of the Mountain of Saint Genevieve, like Harvard versus MIT. Now we understand the liberties taken by the permanent secretary with the documents in his possession.

A number of remarks can be made here about the social context of this affair. The two combatants in this neighborhood quarrel between the Ecole Normale and the Ecole Polytechnique were not Parisians — far from it — but each was closely tied, nevertheless, to a Parisian "temple." The French penchant for centralization appears in all of its splendor. At the same time, for political reasons, Germany was developing numerous universities whose ties with industry did not depend solely on a few brilliant individuals.

Cailletet confirmed his experiment before a competent audience, as was customary at the time. This is what Sainte-Claire Deville called "publish" in his statement to the academy. Note also that Pictet's experiment was quite different from Cailletet's, and not just in procedure. Pictet's apparatus was heavy and complicated — there was no question of transporting it to the academy. This would become more and more often the rule, but it did not immediately eliminate demonstration experiments outside the laboratory or their corollaries, high school laboratory demonstrations.

Such displays could become a bit heroic, as we'll see at the end of this chapter with James Dewar. Dewar even made technical choices, glass versus metal, for example, based on suitability for later public presentations. It is likely that this attitude kept him from being the first to liquefy helium. Everyone knows that an experiment has little chance of succeeding when someone is watching.

Another social aspect of this competition is particularly important to emphasize. Pictet and Cailletet symbolize very well two facets of work in thermodynamics: industry and laboratory. Pictet's company managed the patent process and took out patents in its name. Cailletet, on the other hand, was concerned only about his career. He was completely open to the free use of his results and would allow anyone to copy his apparatus without charge. In 1882, for example, a young Polish scientist named Zygmunt Wroblewski would testify that Cailletet hid nothing about his experiments. Cailletet's laboratory at Chatillon-sur-Seine was surely no less peaceful than the Quiquendone of Dr. Ox, but it has now disappeared under a forest of shrubs because of a quarrel about inheritance.

Researchers at the Ecole Normale in the twentieth century did not break with its tradition of disinterested research. Alfred Kastler never patented the atomic clock nor Raymond Castaing his ionic microprobe. In physics generally this attitude is surprisingly common: Röntgen never took out a patent on his X-ray machines, and Brian Josephson never made the least effort to profit from devices that directly used the properties of superconducting junctions that he had predicted.

Pictet, like Cailletet, had been heavily involved in the problems of liquefaction for a long time, but his goals were industrial. He owned a factory that produced sulfur dioxide (SO_2), known at the time as anhydrous sulfurous acid. This gas provided him with a natural starting point for getting to low temperatures; at ambient pressures SO_2 liquefies at $-10°C$. In fact, Pictet liquefied it in his apparatus by compressing it at room temperature. Then he pumped away some of the gas above the liquid with a vacuum pump. This lowered the pressure of the gas and made the

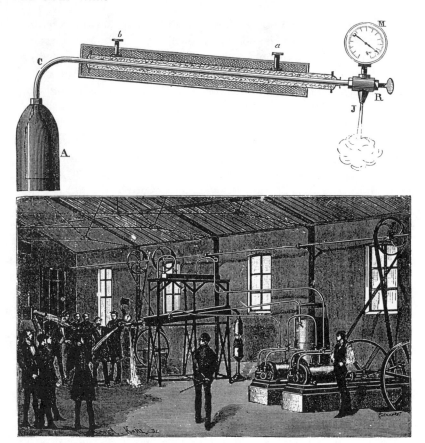

The liquefaction of oxygen by Raoul Pictet.
Top: The oxygen was stored at room temperature at a pressure of 500 atmospheres in a gas bottle A. Compressed gas then passed through the tube C to a heat exchanger, where contact with solid carbon dioxide lowered its temperature, to −130°C according to Pictet. The abrupt release of the pressure in the nozzle J yielded a mist that was interpreted as liquid oxygen.
Bottom: The different elements of the montage are recognizable in this print, where the historic event, complete with high hats and gentlemen's coats, has been staged with particular attention to detail. Note the lantern on the dark side of the atelier that illuminates the jet of mist.

liquid evaporate and so lowered the temperature of the liquid to −65°C. He used this cold SO_2 to cool gaseous CO_2 in another container. Then he compressed the CO_2, pumped on the gas above it, and succeeded in solidifying the CO_2 at a temperature that Pictet estimated as −130°C. In the middle of this solid CO_2 was a four-meter-long copper tube filled with oxygen at 500 atmospheres. Pictet had predicted that under these conditions (500 atmospheres and −130°C) the oxygen in the copper tube would liquefy. In fact, when he let a little escape, he observed a jet containing both gas and liquid. Today we know that the pressure scale that Pictet used to decide the temperature of the CO_2 was not correct. The actual pressure corresponded to a temperature of $-119°C$, barely lower than the critical temperature of oxygen, which we now know is $-118°C$. In addition, thermal contact between solid CO_2 and a copper tube is always poor, so the temperature in the copper must have been a few degrees higher. We are thus forced to conclude, with Martin and Barbara Ruhemann, that surely there was no liquid oxygen in Pictet's tube.[2] Only when the oxygen was released from the tube could it cool below the critical point and liquefy. This is what we now call the *Joule-Thomson effect*.

> In the Joule-Thomson effect a gas contained under pressure escapes through a porous plug that is thermally insulated. Measurements show that the temperature of the escaping jet is different from the temperature of the pressurized gas. For oxygen the temperature of the jet is lower; the same is true for carbon dioxide. This you can see for yourself when you use a fire extinguisher — the foamy blanket of white snow that you produce is solid carbon dioxide.

Cailletet, on the other hand, was content to cool the oxygen pressurized to 200 atmospheres to a temperature of $-29°C$ with liquid sulfur dioxide. When he released the pressure and allowed the oxygen to expand, he liquefied the oxygen, as he well understood. Although Pictet's procedures could produce larger quantities of liquid, Cailletet better understood the results of his experiment. Pictet's idea of using two-stage (cascade) cooling was the one that lasted, because Cailletet's method, a sudden pressure release starting from room temperature, was not very efficient for oxygen, nitrogen, or hydrogen. For every drop of liquid oxygen, he had to get rid of considerable thermal energy, first to cool the gas to the condensation temperature and then actually to liquefy it. The amount of heat that must be eliminated per unit mass in this last step is called the *latent heat of vaporization*. It is very large for oxygen (213 joules per gram). For helium it is much lower (20.9 joules per gram), and so Cailletet's method later turned out to be useful for helium. The high latent heat of nitrogen explains why liquid nitrogen evaporates so slowly. It can be kept for a long time in a container with insulating walls, like the polystyrene in camping coolers, even if the container is open to the air at room temperature.

Before reading Pictet's telegram and Cailletet's sealed letter, Dumas had insisted on reminding his audience of some words of Lavoisier from the previous century:

> Let us think for a moment about what would happen to the different constituents of the earth if its temperature abruptly changed. Suppose, for example, that the earth was suddenly transported to a much hotter region in the solar system, a region where the ambient temperature was much hotter than boiling water. Soon the water and all liquids that vaporize at around the same temperature, and even some metals, would expand and be transformed into aeriform fluids that would become part of the atmosphere.

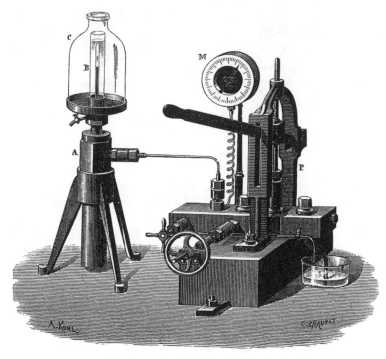

Louis Cailletet's apparatus for gas liquefaction. Versions of this apparatus could still be found in Paris high schools in the 1950s. The gas in a glass tube is liquefied by the pressure of a column of mercury that serves as a piston. This tube is enclosed inside a glass container that is cooled by another liquefied gas. A glass bell jar surrounds this part of the apparatus. Students can thus watch the liquid form and measure the corresponding pressure with the manometer M.

> If, on the other hand, the earth suddenly found itself in a very cold region, say near Jupiter or Saturn, the water that today makes up our rivers and seas, and probably also most of the liquids that we know today, would be transformed into solid mountains . . .
> If this really happened, the air, or at least some of its aeriform constituents, would no doubt cease to exist as invisible fluids, for lack of sufficient heat. They would return to the liquid state, and this transformation would produce new liquids that we can't imagine today.[3]

Now that the obstacle of "permanent gases" was about to be removed, Lavoisier's vision became prophetic; by providing methods of liquefying oxygen, Cailletet and Pictet opened the route to modern cryogenics. (The word was coined about 1878 to describe the science of low-temperature phenomena.) Much later the new liquids that Lavoisier expected but could not imagine turned out to be the isotopes of helium, different forms of helium that would prove to be a new challenge to physicists. Toward the end of the meeting on December 24, someone remarked that, even though we now knew how to transform oxygen from a gas to a liquid, we certainly did not know how to keep it that way. The solution was soon found in Poland.

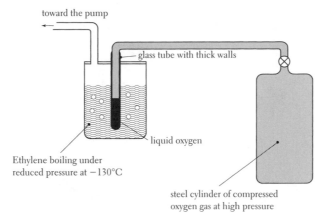

toward the pump

glass tube with thick walls

liquid oxygen

Ethylene boiling under
reduced pressure at −130°C

steel cylinder of compressed
oxygen gas at high pressure

The liquefaction of oxygen by Zygmunt Wroblewski. Wroblewski supplied a proof of liquefaction that both Cailletet and Pictet lacked. By making use of liquid ethylene instead of sulfur dioxide, he cooled the oxygen enough that the abrupt release of pressure yielded an undeniable liquefaction, proved by the observation of liquid in a glass tube.

Zygmunt Florenty Wroblewski, the young Polish physicist mentioned earlier, was a knowledgeable and attentive spectator at Cailletet's experiments. He bought a copy of Cailletet's apparatus from Ducretet, a company that would later be absorbed into the giant Thomson Electric, and brought it back to the Jagellonian University of Kraków, named after the Polish royal dynasty. It was there in the chemistry laboratory that Wroblewski encountered Karol Stanislav Olszewski, who was also in his thirties. Olszewski had been struggling for some years with gases at high pressure and had become exceptionally adept at handling the equipment. These two young men quickly grasped the necessary thermodynamics, and in two months they had found a way to store liquid oxygen. Instead of letting the liquified gas jet escape into the atmosphere, they kept it in a sealed tube. The end of this tube was immersed in an ethylene bath under reduced pressure, which lowered its temperature to −130°C. This experiment proved successful on April 9, 1883, and four days later they sent off their results to the Academy of Sciences in Paris.

A few months later the team broke up. In 1884 Wroblewski and Olszewski each used a technique very similar to Cailletet's and seemed to have found a way to liquefy hydrogen. But it was hard to be sure that liquefaction had taken place because once again there was the problem of storage. Ethylene is highly inflammable; it was clearly dangerous to use. Late one evening in 1892 Wroblewski inadvertently knocked over a kerosene lamp and died in his burning laboratory. A safer way to keep liquid oxygen had to be found.

Cailletet had already thought of a way to get rid of the frost that formed on the glass capillary tubes that he liked to use to demonstrate gas liquefaction. His solution was simple: he would enclose the capillary inside a bigger tube. To get rid of all traces of water vapor, he could put a bit of desiccant between the two glass walls. The enclosed air would then be perfectly dry, so that no water vapor could condense as frost. This kind of setup was tried in the first experiments in Kraków, but it proved to be not quite right. The liquid oxygen was indeed visible, but it bubbled furiously. Better thermal insulation was needed.

Enter James Dewar on center stage. "Little James," the seventh and last son of a Scottish innkeeper, a colorful personality who would accomplish so much for cryogenics, had a bad fall through the ice when he was ten. It took him a long time to get better, and his health remained fragile, but this did not keep him from working right up to his death at eighty-one. The sickly child became marvelously adept with his hands by working with a carpenter who was something of a lutist and fiddle maker. He studied chemistry at Edinburgh and became a professor at Cambridge. He soon left Cambridge, however, for the Royal Institution, the place that was going to become, in a real sense, a theater for his scientific activities. Dewar was naturally eloquent and drawn to the theater. He took great pleasure in the public lectures held every Friday evening, veritable one-man shows of the progress of his work. His junior collaborators were relegated to the background.

As soon as he heard the news that oxygen had been liquefied, Dewar, like Wroblewski, bought one of Cailletet's devices, and he demonstrated it one Friday evening in 1878. This first contact with cryogenics fixed the basic direction of his life's work. But people had to be able to see the results, so it was crucial that the liquefied gas not boil violently. Following Olszewski, he had replaced the dry air in Cailletet's double-walled container with a somewhat more complicated system. The glass tube containing the liquefied gas rested in a container of liquid ethylene. This was placed in a bell jar, so that pumping could lower the pressure of the ethylene vapor.

At this point Dewar remembered that twenty years earlier he had used a vacuum to insulate a calorimeter. The insulation in a calorimeter comes from enclosing a small container (the enclosure), where the measurement is to be carried out, inside a bigger container. As in Cailletet's double-walled system, heat transfer arises above all from collisions of the molecules contained between the two walls, both with each other and with the walls. If this space is evacuated, the primary source of heat transfer disappears. This is what Dewar was able to demonstrate in January 1893. We now call a double-walled container with a vacuum in between the walls a *dewar*.

> A *calorimeter*, as the name suggests, is a device for measuring the heat given off, for example, in a chemical reaction. It is essentially an enclosure that is carefully insulated, so that any heat from the reaction is kept inside and causes a change in the temperature, the measured quantity.

A dewar is essentially a good thermos bottle, made of glass, of course. The only heat transfer comes via radiation from the hot parts to the cold parts. To eliminate

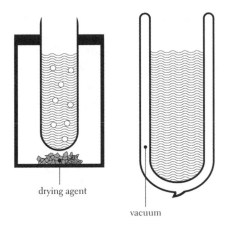

drying agent

vacuum

From Wroblewski to Dewar: improving thermal insulation techniques. Wroblewski used a drying agent to prevent condensation on the walls of his container, but the oxygen bubbled fiercely and made it difficult to see the cold liquid. James Dewar, with his taste for showmanship, tried an intermediate container filled with dry air; this helped the visibility but not the thermal insulation. Both problems were resolved with Dewar's invention of what has come to be called a "dewar," a double-walled glass container with a vacuum in the space between the walls. Dewars with a silver coating on the walls to reflect heat are a standard item in laboratories today, just as "thermos" bottles, which follow the same principles, are handy for picnics and school lunches.

that, Dewar put a coating of silver on the walls inside the vacuum. With no air to dull them, the walls stay shiny and remain good heat reflectors. Such a dewar is also called a *cryostat*. Today the walls are most often made of stainless steel. Stainless steel is a poor heat conductor, it can be easily polished so that it reflects heat well, and it can be readily sealed with solder. The shape of Dewar's container remains the same today as it was in Dewar's day—it's a triumph of design, before the notion of design had even been conceived. Henceforth, anyone who wanted to do cryogenics could not do without a glass blower; Kamerlingh Onnes would remember that.

Dewar was now a competitor in the race toward absolute zero as well as a protagonist in the polemics about priority. Kurt Mendelssohn's excellent book *The Quest for Absolute Zero* gives a good account of these controversies.[4] Dewar (and also M. W. Travers, another Englishman) succeeded in liquefying large quantities of hydrogen in 1898; it could be conserved because of Dewar's dewars. When it was time to find a way to liquefy helium, Dewar insisted on using glass, contrary to the advice of his assistants. This was one of the reasons that helium was liquefied for the first time, not in London but in Leiden, at 4.2K, 4.2C° above absolute zero.

Dewar had just a small amount of helium gas, but he had hoped to see at least a helium mist. The thermal insulation in his apparatus was, after all, one of the best available. But he failed, no doubt because he didn't realize that at liquid helium temperatures even daylight carries a nonnegligible quantity of heat energy. His desperate desire both to see liquid helium and to have it seen did him in. Angry and depressed, he gave up all low-temperature work. Glass had other problems as well; his two assistants who had each lost an eye were testimony to that.

When Dewar left low temperatures behind, efforts in this field of research in England simply stopped. Mendelssohn is sharply critical of Dewar for this and emphasizes how poorly he treated his collaborators. In fact, Dewar did not create a school, but his autocratic behavior was no worse than that of Kamerlingh Onnes, if we can believe the stories of witnesses. Yet their behavior is not the crucial point. Nothing of interest in low-temperature physics happened in Poland after Olzewski. It was just the same in France. Aside from the air liquefiers of Georges Claude, which were an industrial success, low temperature research died out. Perhaps we can attribute this decline to the universities and the academies, which were unable to adapt to a discipline that required theorists, mechanics, and glass blowers to work together. No one realized it yet, but working conditions in science had to change. The era of big science was approaching, and in many ways it was industry that provided the model.

CHAPTER 2

Perpetual Motion?

One man radically changed the way scientists work. He was responsible not only for the discovery of superconductivity but also for initiating developments that characterize modern scientific activity. The man is Heike Kamerlingh Onnes. Born to a family of rich merchants and industrialists, Kamerlingh Onnes studied at Groningen, the Netherlands, and spent several semesters in Heidelberg with G. R. Kirchoff and R. W. Bunsen. After his graduation, in 1879, he taught for several years in Delft before he accepted the chair of physics at the University of Leiden in 1882.

Holland had been an active center for physics since the seventeenth century when Huygens was making his reputation in optics and dynamics, Spinoza was polishing lenses to earn a living, and Descartes was discovering the principles of geometric optics. Leiden itself is an old university town, crisscrossed with canals, with a name that students recognize in the Leiden jar, an early capacitor. The physics laboratories have large bay windows that look out on the street, so a lab worker inside feels like an artisan in his shop; all the passersby stop to watch.

Experimental physicists all fancy themselves artisans. Leiden today gives a rather different impression, but even in those days the laboratory that Kamerlingh Onnes created could never be mistaken for an artisan's shop. He set out his program in his first lecture, like a cabinet minister taking office. His presentation was far more than a simple overview of the scientific program that he wanted to carry out:

> Physics owes its fruitfulness in creating the elements of our material well-being and its enormous influence on our metaphysics to the pure spirit of experimental philosophy. Its primary role in the thoughts and actions of contemporary society can be maintained only if, through observation and experimentation, physics continues to wrest from the unknown ever new territory.
>
> Nevertheless, a large number of institutions are needed for physics to play this role, and they need resources. Both are now woefully insufficient when we consider how important it is to society that physics prosper. As a result, the person who accepts the task of forming future physicists and of managing such institutions must be particularly aggressive in putting forth his ideas about what is really needed to carry out experimental research these days.

Perhaps, like a poet, his work and all his activities are motivated solely by a thirst for truth; to penetrate the nature of matter might be his principal goal in life. Nevertheless, the courage to accept a position that makes it possible to realize these goals must come from the conviction that his activities will be useful only if he follows certain well-defined principles.

What I believe is that quantitative research, establishing relationships between measured phenomena, must be the primary activity in experimental physics. FROM MEASUREMENTS TO KNOWLEDGE (*Door meten tot weten*) is a motto that I want to see engraved on the door of every physics laboratory.

In this kind of presentation it was routine to emphasize the importance of physics; it was merely an expression, after all, of the positivism of the time. What was truly original was the recognition of the importance of the institution. The influence of Kamerlingh Onnes's family background in capital investment is clearly evident. This attitude had important consequences for physics in the future. More than a century later Emilio Segrè, a particle physicist and a student of Fermi, one who spent his life at big particle accelerators, wrote a highly regarded overview of physics in the twentieth century.[1] He was quick to see in this low-temperature laboratory the forerunner of the institutions of "big science":

The passage of physics to a grand scale is usually associated with particle accelerators. This is partly correct but many of the features of future developments appeared earlier: the association of science with engineering, the collective character of the work, the international status of the laboratory, the specialization of laboratories centered on one technique, the division of the personnel into permanent staff and visitors. A laboratory with all these characteristics had been formed by Heike Kamerlingh Onnes (1853– 1926) at the end of the nineteenth century for the study of low-temperature phenomena.

The sterile epistemological dogmatism of *Door meten tot weten*, on the other hand, is surprising. *Meten* (to measure) and *weten* (to know) rhyme in Dutch. It's a new expression of the old idea that there's no science without a measurement. H. B. G. Casimir, a theoretical physicist who knew Leiden very well, sharply criticized this attitude in his memoirs:[2]

Although it is certainly true that quantitative measurements are of great importance, it is a grave error to suppose that the whole of experimental physics can be brought under this heading. We can start measuring only when we know what to measure: qualitative observation has to precede quantitative measurement, and by making experimental arrangements for quantitative measurements we may even eliminate the possibility of new phenomena appearing. (Lenard's set-up was better for certain quantitative studies than Röntgen's, so he didn't discover X-rays.) It is also the task of experimental physics to create new circumstances, so that new phenomena may arise: low temperatures, high pressures, high magnetic fields.

Fortunately, Kamerlingh Onnes often broke his own rule. So here we have a good demonstration that spontaneous philosophizing by scientists is not terribly pro-

The helium liquefaction laboratory in Leiden, shown as it was conceived and realized by Kamerlingh Onnes and as it functioned for decades. Providing this major infrastructure for a basic research laboratory represents one of the most important of Kamerlingh Onnnes's contributions to science. It was a first step toward "big science," in which, for example, outsized infrastructure is necessary for running large particle accelerators and particle detectors.

found. We can add, of course, that the same thing goes for philosophers whose a priori conceptions of science are often equally ill considered.

Kamerlingh Onnes's first concern in Leiden was to provide a substantial infrastructure for his laboratory. It took him ten years to construct the first major technical installation that clearly revealed his determination to break with small-scale operations. This was an elaborate cascade apparatus to provide large quantities of liquid nitrogen, liquid oxygen, and liquid air, which satisfied all the demands of the laboratory for thirty years. Indeed, the liquid oxygen setup was so reliable that it was still in service after Kamerlingh Onnes's death in 1926. Constructed in 1892, three years before the invention of the modern liquefier by C. von Linde, it was based on Pictet's method, which was, however, much less efficient.

While Dewar in London jealously reserved for himself a complete monopoly on the use of his apparatus, Kamerlingh Onnes welcomed anyone who wanted to come and work in Leiden. He went to great and devious lengths to get hold of competent technicians. With a little financial inducement, he managed to steal one of the best glassblowers in Germany, O. Kesselring, who, with the mechanic G. J. Flim, formed a faithful team of assistants. To make better use of their talents, Kamerlingh Onnes started a school for scientific instrument makers and glassblowers right next to his laboratory. Graduates of this school (which still exists) populated physics labs all across Europe. Glassblowing also led to a powerful lighting industry in Holland.

G. J. Flim (*left*) and H. Kamerlingh Onnes in the Leiden laboratory.

The graduate students taking courses were also not forgotten; Kamerlingh Onnes put fifty hours of glassblowing and metalworking into their schedule.

Such an operation does not start working overnight. It took time that also had to be devoted to producing enough electricity, not so much for lighting as for running the pumps and compressors. Throughout this period Kamerlingh Onnes stayed out of the everyday race for quick scientific results, but he was no less a scientist for that. He should not be denigrated as someone whose sole contribution consisted in

H. Kamerlingh Onnes (*left*) and his mentor, J. D. Van der Waals.

introducing industrial methods into science. His vast enterprise succeeded because it was supported by a rigorous scientific program.

Kamerlingh Onnes did not succeed in liquefying hydrogen until 1906, eight years after Dewar, but his procedure yielded much larger quantities. Everything was now ready to attempt to liquefy helium. All the intermediate temperature steps were working so smoothly that he could concentrate on this one final hurdle, which had been the overriding goal of the entire Leiden laboratory for a long time; everything else was secondary. The French physicist Edmond Bauer enjoyed recalling what

Kamerlingh Onnes said to him one day: "You French have too many ideas and too many women. We Dutch have but one idea and one woman." A more scientific anecdote confirms this obsession with helium. Pieter Zeeman, a fellow member of the lab, had to wait until Kamerlingh Onnes was on extended leave before he could carry out his experiments on the effects of a strong magnetic field on optical spectra. His findings, that certain lines split into two or more separate lines, eventually earned him the Nobel Prize.

Kamerlingh Onnes and his competitors soon found that helium would not liquefy in a liquid oxygen bath, even under pressure. They therefore applied the lessons they had learned from liquefying oxygen itself and tried to determine the critical point of helium. Dewar, in 1904, had estimated it to be around 6K by studying the adsorption of helium by activated charcoal cooled by liquid hydrogen. (Adsorption is surface absorption.) Since a gas is more and more easily adsorbed as it gets close to its critical point, his experiment yielded an estimate of the temperature T_c at the critical point, even though it didn't actually get there. The following year Olszewski decided T_c was near 1K, which looked impossible, while Kamerlingh Onnes leaned toward a value of 2K. The liquefaction of helium was thus not a simple technical triumph, but, like that of oxygen before it, it confirmed hypotheses deduced from other experiments.

Helium, the lightest of the gases, had been discovered not long before. An eclipse of the sun on August 18, 1869, a total eclipse above India, provided the first chance to do a spectroscopic analysis of the solar spectrum. J. P. C. Janssen, a Frenchman, and J. N. Lockyer, a Briton, observed an intense yellow line that they could not attribute to either hydrogen or sodium. It had to be a new element, which they called helium, from the Greek *helios*, for "sun." Today we know that helium is abundant in the sun. On earth, however, the situation is completely different. For a quarter of a century afterward no one was convinced that helium even existed. In 1895, however, William Ramsay, who had had some notorious quarrels with Dewar at the Royal Society, analyzed the gas emitted from a sample of heated pitchblende. He was surprised to see not only the characteristic line from argon that he expected but also the yellow line of helium. It turns out that pitchblende contains uranium, which is naturally radioactive, emitting alpha particles, the nuclei of helium atoms. The alphas are stopped in matter after a short distance and attract electrons to form the helium atoms that Ramsay observed.

Fortunately for the authors of this book, helium does exist on earth. It makes up about one part in a hundred thousand of the air that we breathe; it can be found also as gas escaping from oil wells and in most radioactive minerals. Dewar could get it in mineral water from the famous Roman springs in Bath in Somerset (and he once lost his entire stock when a technician left a valve open). Kamerlingh Onnes extracted it from monazite, a natural phosphate of thorium, cerium, and other rare earth minerals. He was proud to recall that, to get the several tons of monazite he needed, he had recourse to the commercial wing of the Netherlands Bureau of Espionage. The contract conditions were excellent—the bureau was run by his brother. The monazite sand came from North Carolina. Once in hand, a squad of young chemists labored in the attic of Kamerlingh Onnes's laboratory heating the minerals, distilling the residue in liquid air and purifying the remaining gas by passing it over cold active charcoal. The result was a stock of 360 liters at the disposal of Kamerlingh Onnes. With so much gas, a thorough study of the isothermal curves could be systematically carried out, so that in 1907 the new results enabled a reevaluation of the critical temperature of helium that placed it near 6K.

In contrast to what transpired later with the discovery of superconductivity, Kamerlingh Onnes himself wrote a detailed account of the first liquefaction, which we summarize here.[3]

All was ready on the tenth of July 1908, to attempt a liquefaction. Flim, the chief mechanic, was on duty to man the liquefier, whose every characteristic had been discussed in detail. The night before, 75 liters of liquid air had been prepared; that very morning, 20 liters of hydrogen had been collected, all to cool the apparatus. The trick in liquefying helium, then and now, is to make absolutely sure that not a trace of air is left in the system. Otherwise, the air will solidify even during the precooling process and block the tubes, especially the escape valve. Three hours after the precooling had started, helium was introduced and began to circulate in the heart of the system. At this point, every step was a new adventure. The only gauge of what was happening was the helium pressure. It was dropping, which meant that the temperature was also dropping. Suddenly, however, the pressure no longer wanted to budge; it seemed completely stuck. Nothing was happening, and it was already 7:30 p.m. Not one drop of liquid hydrogen was left; the experiment looked dead. People in the lab knew that a big experiment was under way, and they came by to hear the news. One of them suggested that, after all, it would be pretty difficult to see liquid helium. The fact that the thermometer seemed stuck could just as well mean it was immersed in liquid helium that was boiling away. In that case the pressure wouldn't budge either. How about trying to illuminate the tube of possibly liquid helium from below? No sooner said than done—what a relief! In the glass tube a thin black line revealed the level of the liquid as it reflected the light. A half-cup of helium had been liquefied. An attempt was made right away to solidify the helium by pumping out the tube to cool the liquid further, but it was not successful.

Liquid helium proved to be more complicated than any ordinary liquid. Kamerlingh Onnes had wanted to approach absolute zero in one fell swoop. Instead, he opened myriad new paths for research in the low-temperature world. For years he tried to solidify helium, without success; it was liquid helium that would make him famous.

Now we can begin the story of the Dutch phantom who discovered superconductivity. Contrary to what you read in physics books, superconductivity was not discovered by Kamerlingh Onnes.

With the liquefaction of helium Leiden had demonstrated its superiority in both pure science and in cryogenics. It had taken a quarter of a century to reach this point. Liquid helium had been extracted under conditions that had little to do with

Door meten tot weten. Now that it was available, a whole series of experiments became possible. The rhythm of work in Leiden was soon fixed once and for all by Kamerlingh Onnes, and it continued for a long time even after his death. Helium was liquefied once a week with Flim in charge. By the end of the morning a physicist would find a cryostat full of helium ready for him to use as he pleased.

Kamerlingh Onnes decided that the entire laboratory would embark on a systematic program of measurements of the properties of matter at low temperatures. Measurements of the resistivity of metals as a function of temperature had a significant role to play in this program. As long as Flim was around to take charge of anything that involved liquid helium, the experiments were rather simple to carry out, and they were crucial in achieving a theoretical understanding of the electrical resistance of metals. Previous measurements had already shown that resistance fell with decreasing temperature. Kamerlingh Onnes thought that this effect could be related to interactions of electrons with metallic ions that vibrated due to thermal agitation. Cooling would reduce the vibrations, and so the resistance would also diminish. Two physicists were assigned to this program. The first, Gilles Holst, had a degree in physics and mathematics from the Polytechnic Institute of Zurich (ETH), where Einstein had studied. Holst had come as a lab assistant to start work on his doctoral thesis. The other physicist, C. Dorsman, already had his doctorate.

Dewar, meanwhile, had already attacked the temperature variation of resistivity with his liquid hydrogen. He had shown that the resistivity of gold and platinum wires decreased with lower temperatures until it reached a minimum value. He thought that the resistivity might rise again if later experiments could lower the temperature further. In Leiden, however, the first experiments with liquid helium by Kamerlingh Onnes's group reached absolute temperatures ten times lower and revealed a totally different scenario.

Resistivity was traditionally measured with a Wheatstone bridge, a classic and rather simple arrangement of resistors valued for its precision. Precision also seemed to demand that all impurities in the metal be eliminated; they were quite rightly suspected of playing an important role in resistivity. For example, gold is an excellent conductor at room temperature and can be purified easily because it has a low melting point. It has a lower resistance at low temperatures than platinum, another good conductor, which has a higher melting point. In 1911 mercury was the metal chosen to pursue this purification one step further; it was readily available, since it was routinely distilled for use in thermometers. A precise measurement also required good electrical contact between the metal and the wires of the Wheatstone bridge. With mercury that was easy. The wires were simply dipped into the liquid metal, which solidified as the temperature was lowered. It was Gilles Holst who carried out all the measurements.

The result was a complete surprise. Instead of a smooth decrease toward zero, the resistivity of the mercury abruptly went into free fall at 4.23K; within three hundredths of a degree the resistance was absolutely undetectable. Superconductivity had been discovered.

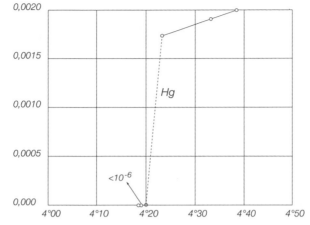

The superconducting transition in mercury. The graph is similar to the one shown in the original article in which Kamerlingh Onnes announced the discovery of superconductivity. The values of the resistance measured by Holst are plotted vertically versus the temperature (on the Kelvin scale) horizontally. At a temperature slightly above 4.20K the resistance falls dramatically from 0.002 ohms to a value too small to be measured, less than 10^{-6} ohms. The dashed line indicates that Holst found it impossible to make any measurements in the region between the resistive state and the superconducting state.

The results of the measurements were published in *The Communications of the Leiden Laboratory*, a series that Kamerlingh Onnes began not long after the laboratory was created. In that era the *Communications* was the primary source of scientific information about cryogenics. Kamerlingh Onnes was the sole author of the paper that proclaimed these results; he acknowledged Dorsman and "Mr. G. Holst, who carefully carried out all the measurements with the Wheatstone bridge." The formula was similar to the detailed account of the first measurements of resistivity, which concludes with the following paragraph: "I acknowledge with gratitude my debt to Dr. C. Dorsman for his intelligent assistance during this entire investigation and to Mr. G. Holst, who carefully carried out all the measurements with the Wheatstone bridge."

These days Gilles Holst would have been coauthor of the papers. Kamerlingh Onnes may not even have been present at the crucial moment. It is true that superconductivity required no new technical equipment. The Wheatstone bridge was commonplace, even if the actual precision achieved, one part in a hundred thousand, revealed the watchmaker's carefulness of Kamerlingh Onnes. In addition, the program planned by Kamerlingh Onnes would inevitably have led to the discovery of superconductivity; any physicist could have carried out the measurements. Finally, we note that it was not unusual at the time to set young collaborators adrift like this. The problem, however, is not just that Holst was not an author; his name seems to have simply disappeared. Mendelssohn, one of the best references on the history of low-temperature physics, does not mention him.[4] Kamerlingh Onnes, who left us with a detailed account of the liquefaction of helium, wrote nothing at all comparable about the birth of superconductivity. For the liquefaction he even cited his colleague who happened by during the experiment and suggested that the cryostat be illuminated from below. Surely, someone would have thought of this before too long anyway. For the liquefaction of helium we have essentially a journal, with spirit and life. For the birth of superconductivity we have but a summary account, cold and dry.

Fifty years later, when Mendelssohn, himself a well-known low-temperature physicist, was comparing Dewar and Kamerlingh Onnes, his criteria were not strictly scientific. Dewar was always in the wrong, and, by contrast, Kamerlingh Onnes always looked good. Dewar, he said, never founded a school that survived him, and he never allowed his assistants to be full partners in his work, which he alone authored. Kamerlingh Onnes, on the other hand, credited his collaborators' contributions. To us the difference between the two men seems rather less striking than Mendelssohn suggests. While Kamerlingh Onnes did create a remarkable institution, the papers from Leiden show that he was willing to share authorship and credit only with well-known physicists. According to Casimir's recollection of the eyewitness testimony of those who knew him well, Kamerlingh Onnes was just as much a scientific despot as Dewar. This, too, was hardly unusual at the time.

According to Casimir, Kamerlingh Onnes's dictatorial attitudes had more to do with assumptions of class privilege than scientific elitism. His social prejudice extended even to the renowned theorist Hendrik Antoon Lorentz, a colleague in the Leiden lab. Because of his modest origins, Lorentz too was subject to the silent hostility of Kamerlingh Onnes. Lorentz's daughter,[5] herself a physicist who later married another physicist, Wander de Haas, tells how Kamerlingh Onnes refused her father a few extra square meters that he had been promised when the laboratory was expanded in 1906. Kamerlingh Onnes had the same attitude toward Holst. Two years after the discovery of superconductivity, his articles continued to minimize Holst's contributions: "This concludes the H series of my experiments with helium. I want to express my gratitude to Mr. G. Holst, an assistant in the physics laboratory, for the devotion with which he has helped me, and to Mr. Flim, the head of the technical division of the cryogenics laboratory, and Mr. Kesselring, the glass blower at the laboratory, for their extensive help in setting up the experiments and building apparatus."

Holst, the son of the director of a naval shipyard, apparently was not entirely satisfied with this kind of acknowledgment, and he left the laboratory soon after. A Dutch company based in Eindhoven that was doing well in the lighting business had a factory in the south and had always wanted to create a laboratory next door, and the lab needed a director. The ancestors of the owners of the company had connections to the family of Karl Marx, and both families had been converts to Christianity. The company still exists—Philips, the global electronics giant. Gilles Holst was the first head of its laboratory. He left Leiden in 1913, the same year that Kamerlingh Onnes received the Nobel Prize. The following year several papers were published with Holst as coauthor. Casimir recounts in his memoirs that, even when Holst was director of the Philips laboratories, he would never dare ask to be a coauthor on a paper describing work in which he had participated unless it was he himself who had come up with the idea in the first place. In spite of everything, at the end of his life Kamerlingh Onnes did give credit to his colleague. In a confidential letter in 1926 to the Royal Academy of the Netherlands, he emphasized Holst's role in the discovery of superconductivity and recommended that Gilles Holst be elected to the academy.

Physicists, like politicians, like to name things after famous people. In magnetism, for example, the temperature below which a material becomes ferromagnetic is called the Curie-Weiss temperature. But names cannot be taken lightly; everyone must agree, or things can become chaotic. And so it is that in optics what the Americans call Snell's Law becomes, in French, the Law of Descartes. A cryogenicist recounts how, several years ago, an international committee took up the task of rationalizing the nomenclature of low-temperature physics. The members noted with alarm that the critical temperature for superconductivity — the temperature below which a metal becomes superconducting — was called quite simply the *critical temperature*. Expecting unanimous agreement, one member of the committee proposed that, in analogy with the term in magnetism, the critical temperature for superconductivity be named the "temperature of Kamerlingh Onnes." The representative of Holland, so it is said, was violently opposed.

When they weren't having these family quarrels, the physicists had some real work to do, thinking about what a superconductor might possibly be. There was a touch of irony here. According to classical thermodynamics, the approach to absolute zero was an approach to a state of total immobility. Nothing moved, a situation conceivable only for solids. This idea was now turned upside down. For one thing helium couldn't be solidified, and now the physicists found themselves faced with something completely new and unforeseen, a metal with no detectable electrical resistance. Once started, a current would presumably flow in such a metal forever. Kamerlingh Onnes, in fact, tried just that experiment, with a current flowing in a superconducting lead ring. Currents produce magnetic fields, so he could look at a compass and figure out whether the current was changing. He saw no change. The title of the article he wrote about this experiment conveys his conclusion exactly: "The Imitation of a Molecular Amperian Current or Permanent Magnet in a Superconductor."[6] In a letter to Lorentz the renowned physicist Paul Ehrenfest expressed the feelings brought on by this phenomenon: "It's very strange to watch the effect of these 'permanent' currents on a magnetized needle. I can almost feel in a tangible way how the ring of electrons is turning, turning, turning in the wire, slowly, almost without disturbance."[7]

Indeed, there was plenty there to get people excited, so much that it led to a Nobel Prize for Kamerlingh Onnes, the first Nobel related to superconductivity, even if the award was for liquefying helium. In his Nobel lecture Kamerlingh Onnes introduced the new word (which had previously been *supraconductability*): "At 4.2K mercury transforms into a new state. Because of its unusual electrical properties, this should be called the superconducting state."[8]

In addition to infinite electrical conductivity, another phenomenon was also observed in Leiden at this time. A sample of metal was cooled below its superconducting transition and placed in a magnetic field. For sufficiently strong magnetic fields (about 0.04 teslas, the *critical magnetic field* for mercury) the superconductivity suddenly disappeared, and the metal became "normal" once again.

Because current-carrying wires create their own magnetic fields — a superconductor could become normal even when it was not in an external magnetic field — Kamerlingh Onnes originally imagined that superconducting wires could be used to make magnets that required no energy input. So it was an intense disappointment when he found that, as he raised the current in the superconductor, the superconductivity suddenly disappeared when the magnetic field was only a few hundredths of a tesla.

> The first explanation for this behavior that might come to mind is that the wire is getting hot, the way a normal wire does when a current is flowing, because of its resistance. But this idea won't work, because the superconducting wire has no resistance at all. What happens, in fact, is that the current in the wire produces a magnetic field at the surface of the wire. When the current is increased sufficiently that this magnetic field reaches a critical value, the superconductor becomes normal. This current is labeled the *critical current* for the superconductor.

The Leiden laboratory continued to be active with the start of World War I in 1914; in fact, it was the only laboratory to do so. When shipping stopped, however, the pace slowed until the end of the war. No more ships at the dock, no more monazite for Kamerlingh Onnes's lab as a source for helium. New supplies could no longer make up for the routine losses of the precious gas. The time between liquefactions became longer and longer. The history of low-temperature physics during the war can be summarized by a brief recounting of several theoretical contributions, in fact the first in this area. In 1915 the English physicist J. J. Thomson maintained that superconductivity was "one more piece of evidence, this time fatal, against the theory that metallic conduction is due to the presence of free electrons which drift under the influence of an electric field."[9]

For Thomson superconductivity confirmed his corpuscular theory of matter, a point of view that was still widely accepted in the early twentieth century. Indeed, Einstein had to battle against it during an informal seminar in Leiden in 1920. While Thomson's position seems quaint today, it was not so far-fetched. He was wrong in arguing against the idea of free electrons in metals, but his proposal for explaining superconductivity was similar to Pierre Weiss's explanation for the appearance of ferromagnetism below a certain temperature. The desire to establish parallels between superconductivity and ferromagnetism was strong in the period between the two world wars.

That same year (1915) F. A. Lindemann suggested that electrons form a lattice that slides without disturbance through the crystalline lattice of a metal. In 1916 and 1917 C. Benedicks and P. W. Bridgman wrote a series of papers defending the hypothesis that it is the valence electrons, the outer electrons, which jump from one atom to the next.[10] At low temperatures certain electron orbits would become tangent to one another, so that the electrons could easily move along extended molecular chains and the conductor would become a superconductor.

Once peace returned, scientific activity in Leiden once more took up the rhythm determined by Kamerlingh Onnes, as if there had been only a small intermission. In addition to Lorentz, we must add to the list of members of the laboratory Paul Ehrenfest, a Russian emigré and friend of Einstein, and Peter Debye, who in 1912 was already using quanta to calculate the specific heat of metals.

> *Specific heat* is the name for the heat energy that must be supplied to a given amount of material to raise its temperature by one degree. In standard units the specific heat is expressed in joules per kilogram per degree Celsius. More details are described in chapter 3.

In spite of this abundance of talent, rather little theoretical work was devoted to superconductivity during this period. It was quantum theory that was drawing theoretical attention. Elementary notions that we take for granted today were hotly debated, even by physicists who inspire the greatest respect. It took forty-six years before the puzzle of superconductivity was resolved. The basic difficulty was that no one really knew how to explain conductivity in an ordinary metal.

CHAPTER 3

Metals and Theories

For hundreds of years a purely descriptive definition of a metal seemed satisfactory. At the end of the nineteenth century (and often even today) a metal was defined by its luster, by its high electrical and heat conductivity, and by its mechanical properties. Yet it's not uncommon to find materials that are hard and ductile or which shine like a metal or even which conduct heat well but are not metals. Diamonds are good heat conductors but electrical insulators; graphite is a fairly good electrical conductor, but it's a heat insulator. Neither one is a metal. Pyrite is called "fools' gold" because it shines so brightly, but it doesn't conduct electricity. And so scientists today have a more microscopic definition: a metal must contain electrons that move around freely, even at the lowest possible temperatures.

Apart from mercury, metals are solids — even crystalline — at room temperature. Their atoms are arranged in an ordered lattice, forming repetitive patterns with characteristic geometric structures. A piece of copper or iron, however, does not look like a crystal you would see in a geology exhibit. This is because copper is not one large single crystal — it's made of myriad tiny crystals embedded in one other, invisible to the eye but easily seen with a microscope.

The (negatively charged) electrons and the (positively charged) ions in the lattice each play a role in defining the properties of metals. Understanding the interplay between them ultimately led to an understanding of superconductivity in the latter half of the twentieth century. It came only after years of careful experimentation on the properties of metals and slow progress in the theoretical explanation of these properties.

Abundant experimental data available well before Kamerlingh Onnes's time showed that two crucial transport properties of metals are closely related. Metals that are good (poor) heat conductors are also good (poor) electrical conductors. Silver is an excellent electrical conductor; a silver spoon in a cup of tea quickly becomes too hot to handle. Not so for the iron handle of the perfect frying pan for your omelette, nor would you want to have iron wires carrying electricity to your stove. The relationship between the electrical conductivity and the heat conductivity of metals is quantified in the *Wiedemann-Franz law*, enunciated by two experimenters, Gustav Wiedemann and Rudolf Franz, in 1853: The ratio of the electrical conductivity to the heat conductivity is the same for all metals at a given temperature. This law was experimentally well confirmed, but it posed a challenging problem for theorists. Could a single theory describe the two properties so accurately that it would correctly predict the value of their ratio? The response requires a close examination of these two seemingly rather different properties.

Electrical conduction involves current and voltage and, of course, the conductor itself. We say that there is a current in a certain direction when particles with positive charge are moving in that direction or when negatively charged particles are moving in the opposite direction. A force is needed to get the charges moving; that's where the voltage comes in. A voltage difference, provided by a battery connected to the two ends of a wire, for example, will yield a force on the electrons in the wire. A twelve-volt battery will produce a bigger force and more current than a six-volt battery. Finally, the conductor is nothing but the material in which the charges move.

> Think about the federal budget. When the government wants to enrich its coffers, it can go about it in two ways: increase taxes or cut down on expenses. Taxes are like the positive charges and expenses like the negative charges. Increasing taxes has the same effect on the coffers of the state as decreasing expenses.

Now consider what happens in the wire. Charged particles leave the battery at one end of the wire, flow through the wire, and enter the battery at the other end. The moving charges constitute a current, measured in amperes. The more charged particles available and the faster they move, the bigger the current. A current of one ampere, a quite ordinary current in a light bulb, corresponds to the passage of billions of electrons per second past any point in the wire. The size of the current depends on the voltage, on the length and size of the wire, and on the particular metal the wire is made of. The ratio of the voltage across a wire to the current is called the *resistance* of the wire (measured in ohms); it's a measure of the obstacles to the flow of charged particles in the wire. The inverse of resistance is called *conductance*; a wire with high resistance has low conductance. A current of one ampere will exist in a wire whose resistance is one ohm if the voltage across the wire is one volt. These units, which have become commonplace words, are all named after famous physicists, Georg Ohm, André-Marie Ampère, and Alessandro Volta.

The resistance of a particular wire depends on the material it's made of and on its geometry. A long wire has more resistance than a short wire of the same material; thin wires have higher resistance than thick wires—these are "geometrical" effects. A short iron wire has a higher resistance than a short copper wire. That's because iron has a higher resistivity than copper. The resistivity (and its inverse, the conductivity) characterizes the resistance of a given material independent of the geometry.

TABLE 3.1
Resistivity and Conductivity for Various Materials

MATERIAL	CONDUCTIVITY ($\Omega^{-1}\,M^{-1}$)	RESISTIVITY ($\Omega\,M$)
Silver	6.14×10^7	1.62×10^{-8}
Copper	5.88×10^7	1.70×10^{-8}
Aluminum	3.54×10^7	2.82×10^{-8}
Silicon	1.6×10^{-5}	$62,500$
Glass	10^{-10} to 10^{-14}	10^{10} to 10^{14}
Quartz	1.33×10^{-18}	7.51×10^{17}

The vocabulary needed to discuss heat conductivity is more familiar. Let's return to the silver spoon in the teacup. After a few minutes it becomes hot over its entire length, even the part above the liquid; thermal energy travels from the hot end toward the cold end. The difference in temperature plays the role of the voltage difference, the flow of thermal energy resembles the current, and the spoon itself is the (heat) conductor. Thermal conductivity is defined just like electrical conductivity, but the units are not named after physicists.

By 1900 it seemed that researchers had all the ingredients they needed for a theory that would explain the properties of solids, and particularly metals. Physics and chemistry had made spectacular progress over the last hundred years. Electromagnetic theory had been developed by James Clerk Maxwell and statistical mechanics by Ludwig Boltzmann. The atomic nature of matter had been confirmed, and J. J. Thomson had discovered the electron in 1897.

The German physicist Paul Drude proposed the first theory of metals that seemed reasonable in 1900. His theory provided an explanation of the Wiedemann-Franz law that came astonishingly close to predicting the experimentally measured value of the ratio of thermal and electrical conductivities. The theory is sufficiently simple that it's possible to sketch it here.

Drude thought of metals as solids with atoms in orderly rows and electrons moving about freely. When a voltage difference is applied to the metal, all the electrons feel the same force and move in the same direction. This movement constitutes the current in the metal. Drude's idea of current in a metal can be naively illustrated by a pastoral scene. A shepherd, far from his sheep, finds that they have spread out far and wide over a field filled with a random assortment of bushes, thickets, and patches of prickly weeds. When the shepherd calls his flock, the sheep all start off in his direction, but each one tries to avoid the obstacles in its path, constantly twisting and turning and starting off anew in the right direction. The average speed of our flock, not so high to begin with, is noticeably reduced by the briars and brambles they encounter. The metaphor is clear: the sheep are the carriers of electric charge, the voice of the shepherd is the voltage, and the poorly maintained pasture is the conductor. The obstacles that slow down the sheep create the resistance of the conductor. To translate our metaphor into a theory of electrical conduction, we would have to decide what it is in a length of copper wire that corresponds to the sheep and what to the briar patches. Drude declared that the electrons carry the current, and the atoms in the lattice are responsible for the resistance.

Now we must consider how the electrons and the atoms participate in thermal conductivity, the other component of the Wiedemann-Franz ratio. The thermal energy, the heat, in all solids is the energy associated with the vibrations of all the atoms around their equilibrium positions. The equilibrium position of each atom is determined by the chemical bonds linking it to its neighboring atoms. A metallic crystal is like a lattice of balls (the atoms) attached by springs to their neighbors, so that only small oscillations of the balls are possible. When one ball is disturbed, the springs transmit the disturbance to the neighboring balls, which then start to vibrate.

The atoms in the crystal are like swings hung in a long row from a crossbar. If the crossbar isn't tightened properly at the ends, it will twist a bit when a swing moves and transmit the movement of one swing to all the other swings in the row. After an instant most of the swings are moving, and the swing that started it all might even be momentarily at rest. An attentive observer would notice that the movement of the different swings is not random. They might all be oscillating together, or maybe one of a pair will be going forward and the other backward. With many swings there are many different possibilities of arranging the motion of each swing relative to all the others. In fact, there are just as many configurations possible as there are swings. If we apply these ideas to a metal, we find that the number of possible modes of oscillation is enormous. A gram of copper can be disturbed in billions upon billions of different ways, and each way corresponds to a different microscopic arrangement of the atoms, a different way of being hot.

So, the vibrations of the atoms are themselves the heat, the thermal energy of the solid. The coupling between the atoms via the springlike force allows this energy to propagate across the solid. We know, however, from the Wiedemann-Franz law, that there is a relation between thermal conductivity and electrical conductivity, and Drude correctly proposed that the latter involves electrons. Our model of balls and springs doesn't give the electrons much of a role in thermal conductivity, so we have to change the model. Think about the swings again. Suppose children sitting on neighboring swings face each other, that is, they face in opposite directions. Each is holding a rather heavy ball, and they can have fun throwing the balls back and forth to each other. If a girl at rest throws a ball forward, she'll recoil backward, and her swing will start to move; the same thing will happen to the boy who catches the ball. All the swings can start oscillating this way, probably much more quickly than before. Most of the energy of motion is still stored in the swings, though now some energy is contained also in the heavy balls. Nevertheless, the transfer of energy from one swing to another depends almost entirely on the balls; the crossbar is much less important. This analogy aptly describes Drude's concept of what happens in metals:

itinerant electrons transport thermal energy by colliding with atoms just as they transport electricity.

Drude translated these ideas into a quantitative model. He used a few equations from classical mechanics as well as some hypotheses about the number of electrons moving about and the kinds of collisions they undergo. Drude's model is so straight-forward that it continues to be taught in elementary physics courses. Students who don't get beyond this level thus have a remarkably clear idea of how electrical and thermal conduction take place in a metal. Unfortunately, the model is wrong.

Drude's model was an immediate success when it was proposed in 1900. The predicted value of the Wiedemann-Franz ratio of thermal to electrical conductivity was amazingly close to the experimental value. When this ratio is multiplied by the absolute temperature, the result is called the Lorenz number. This Lorenz, L. V. Lorenz, has nothing to do with H. A. Lorentz of Leiden, whom we will meet again shortly (note the difference in the spelling of their names). The Drude prediction for the Lorenz number was 2.48×10^{-13} (in standard units), and the agreement with the experimental value of 2.72×10^{-13} seemed too good to be fortuitous.

Nevertheless, not everyone was satisfied. The model seemed too crude, which is not surprising, given that it was introduced only three years after the electron was discovered. For example, the free electrons constituted a gas, a fact that the calcula-tions essentially neglected. The classical statistical model of Maxwell and Boltz-mann had yielded some remarkably good results in applications to the theory of gases, and it was expected to provide a better approach to some of the issues involved in metals as well. This tempted the distinguished Lorentz to redo the calculations of Drude with much more sophisticated mathematics.[1] This he did in 1904–1905 in Leiden, treating the collisions between electrons and atoms in much more rigorous fashion. Like Drude, Lorentz found a universal ratio of thermal and electrical con-ductivities, but there was an important difference. Lorentz's new, supposedly better founded, prediction of the Lorenz number was in poor agreement with experiment. His predictions for the temperature variation of thermal and electrical conductivi-ties were also in flagrant disagreement with experiment.

Clearly, something fundamental was missing. Maxwell-Boltzmann theory, which had so successfully described the behavior of gases, apparently did not hold for metals. Once again the theoretical models developed during the nineteenth century were shown to be lacking, and once again, just as in electromagnetism, Lorentz's work proved to be the turning point between classical physics and the new physics just beginning. There is no shortage of epistemological theories about the formation and refutation of ideas in physics. Lorentz's calculations showed that Drude's model had to be rejected, but this important counter-intuitive example in no way illustrates any of the refutation schemes that have been proposed by various modern philosophers of science.

Lorentz's failure stopped most theoretical attempts to understand metals, while work on nonmetallic solids continued. In 1912 Petrus Debye worked out a model of

atomic vibrations and thermal energy in crystals. Debye represented a new generation of Dutch physicists; Kamerlingh Onnes was already a professor when Debye
was born. Debye's model was an elaboration of a key idea proposed by Einstein three
years earlier. Einstein wanted to show that the quantum hypothesis could be successfully applied to the mechanical vibrations of atoms, just as it had previously been
applied to light and the photoelectric effect. Einstein had thus applied the Planck
relation $E = h\nu$ to the vibrations of a solid. Here E designates a *quantum* of vibrational energy, ν is the frequency of the vibration, and h is Planck's constant. Today
a quantum of vibrational energy (the smallest unit possible) is called a *phonon*, in
analogy with the quantum of electromagnetic energy, which is called a *photon*. In
Einstein's model all the atoms vibrated independently with the same frequency.

Planck had proposed his relation to explain the spectrum of radiation emitted by a hot object, a so-called blackbody. As Planck recalled in his autobiography, he was "not only indifferent, but to a certain extent even hostile to
the atomic theory."[2] In spite of these strong misgivings, he introduced the
quantum hypothesis and so eliminated this "dark cloud" — the puzzle of
blackbody radiation — from the physicist's sky.

Debye's model is still used. Rather than considering each atom individually,
Debye considered the solid as a whole. Just as in our swing analogy, the solid could
have a large number of possible vibrational frequencies, in fact, three times the
number of atoms in the crystal. We would say today that Debye considered phonons
of many different frequencies, whereas Einstein's phonons had a single frequency.
Debye's model was able to predict values of the *specific heat* of solids at low temperatures in good agreement with experiment.

The specific heat of a solid is a measure of the energy that must be supplied
to raise its temperature. Measurements of specific heats had been carried
out for a long time. The Law of Dulong and Petit had been known since
1820. It describes the experimental fact that the specific heat is the same
for all solids; it takes the same amount of thermal energy to raise the temperature of N atoms of copper and of iron, for example, by one degree. This
discovery had even served as an argument in favor of atomic theory at the
time. The experimental value of this universal specific heat at room temperatures was well known and had been explained nicely by the statistical
mechanics of Maxwell and Boltzmann. This value also appeared naturally
at very high temperatures in the Einstein and Debye models. At low
temperatures the situation is more complicated. Einstein was the first to
achieve a rough description of the specific heat at low temperatures, and
Debye made a significant improvement.

Debye's model also included a characteristic temperature for each crystal, later called the *Debye temperature*, at which all possible vibrational modes of a crystal can be excited. The Debye temperature is high if the bonds between the atoms are very strong, that is, if the substance is very hard, because it takes considerable heat to excite all the vibrations. The hardest substance we know, diamond, has a Debye temperature of 2230K. On the other hand, lead, a soft substance, has a Debye temperature of only 105K.

This digression about specific heat is important for three reasons. First of all, it illustrates the significance of the vibrational modes of a crystal, which will be necessary to describe superconductivity. It demonstrates the importance of low-temperature measurements to researchers at the turn of the century. Finally, it allows us to understand the most serious criticism of the Drude-Lorentz model of conductivity. In their model a gas of electrons circulates freely among the atoms of a metal. To make the model agree with experiment, each atom has to provide one or more freely moving electrons. Maxwell-Boltzmann statistics prescribed that this electron gas had to make a contribution to the specific heat that was roughly equal to that of the atoms themselves. But Debye's model explained specific heat without including any contribution from the electrons. The electron gas just wasn't there — or so it seemed.

The impasse was total. Some crucial pieces were missing from the puzzle, and the agreement between the experimental value of the Lorenz number and Drude's theory was probably fortuitous. Despite occasional successes, this situation hardly changed for twenty-five years. Even the physics elite who attended the Solvay conference in Brussels, Belgium, in 1924 could do nothing but recognize that the problem was serious. (This extraordinary meeting was one of a series that were financed by Ernest Solvay, the inventor of a process for making baking soda. They were occasions for a small number of the most famous physicists and the most talented newcomers to describe their discoveries, each trying to outdo the other.)

Between 1928 and 1933, however, everything changed in physics. Three names and three cities are associated with these prodigious advances, particularly in the physics of metals: Arnold Sommerfeld in Munich, Werner Heisenberg in Leipzig, and Wolfgang Pauli in Zurich. These were the leaders, but there are many others who also deserve mention—Hans Bethe, Felix Bloch, Léon Brillouin, Rudolf Peierls, Lev Landau, John Slater, and A. H. Wilson. In five years, in an atmosphere of intense excitement, they cleared up most of the problems that had plagued solid-state physics, and particularly metals, for the last century.

In the early 1920s the chief obstacle to all progress in the theory of metals was in understanding the role of the electrons. Their behavior presented a real paradox. Electrons seemed necessary to explain most metallic properties, but few were found whenever an experiment was designed to count them. The specific heat that we just talked about is a good example, as is magnetism.

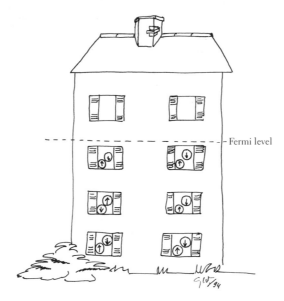

The Pauli exclusion principle. The Pauli principle says that it is impossible for two or more particles of half-integer spin to occupy exactly the same quantum state simultaneously. Two — but no more than two — electrons, which have spin ½, can occupy the same energy state provided their spins point in opposite directions (so their quantum states are not identical). The energy states of the system are represented in this drawing by apartments in a tall building. At a temperature of 0K the electrons occupy the lowest energy states (the lowest apartments) possible, until they have all been housed. The energy of the highest filled state is called the *Fermi energy*.

No doubt it was just this situation that provoked Wolfgang Pauli, an Austrian born at the beginning of the twentieth century, to ponder the nature of this gas of electrons that was thought to fill all metals. In 1925 Pauli had formulated his famous *exclusion principle*, which forbids two identical particles in one atom to occupy exactly the same quantum state. Particles in the same quantum state would have the same energy. A quantum state of an atomic particle is something like a person's home address, specifying the country, state, city, street and street number, and, for an apartment building, the floor and room number. If people were subject to the Pauli principle, two persons could not occupy the same "state" — that is, the same apartment. Electrons act as if they are spinning like the earth on its axis. The exclusion law for electrons and other particles with spin was a little less rigid — two electrons with opposite spins may have all their other quantum numbers identical. As long as one electron appears to spin clockwise and the other counterclockwise, they have the right to coexist in the same state.

Working independently of each other, two other bright young stars of the physics world — Enrico Fermi, born in Rome in 1901, and Paul Dirac, born in Bristol, England, in 1902 — applied this exclusion principle to a gas of electrically neutral "electrons." Without their negative charge, these purely imaginary electrons no longer repel one another; they could share space in a small box without having the box explode. The new quantum mechanics of Erwin Schrödinger made it possible to calculate the properties of all the quantum states allowed for such particles confined to a "box" (like a strip of copper). He showed how to classify these states by increasing energy as if they were apartments in a giant building, numbering each apartment starting at the ground floor and going up, one story after another, as high as necessary. Fermi and Dirac had only to put the electrons in the apartments following the Pauli exclusion principle.

If N persons wanted to live in such an apartment building, which we'll suppose has no elevator, the Pauli principle would suggest that one couple goes into apartment 1 on the ground floor, then another couple into apartment 2, and so on, until the ground floor is filled. Without an elevator no one would want to go to a higher floor as long as the ground floor had a vacancy. The second floor would then be filled in the same way, and so on, until all the applicants were satisfied. The lowest N/2 apartments would be occupied; there are empty apartments on higher floors. This analogy corresponds to the states of an electron gas at 0K confined in a box according to the distribution law suggested by Fermi and Dirac. The absence of an elevator corresponds to the need for more energy to go to a higher state. The lowest N/2 states are each occupied by two electrons of opposite spin. The filled state of highest energy is called the *Fermi level*; its energy is labeled E_F. The energy of this state depends on the volume of the box and the number N of electrons. States with energies higher than E_F are all empty at 0K.

In a metal the energy of the electrons is essentially entirely kinetic energy, energy of motion. What the Fermi-Dirac model says is that even at 0K the electrons in the highest filled states, up close to the Fermi level, have large kinetic energies. Contrary to the simple classical idea that nothing moves at 0K, such electrons can be speeding around the metal at 10,000 kilometers per second or more. The classical picture is based on the Maxwell-Boltzmann theory, in which the average kinetic energy is proportional to the temperature, namely, $3kT/2$; k is a constant named, not surprisingly, the Boltzmann constant. At 0K the average kinetic energy would thus be zero and all the electrons would be at rest. As the temperature is increased, all the electrons would be disturbed and start to move; their average energy would increase linearly with T. Classical electrons don't obey the exclusion principle.

The electrons in a metal, however, don't follow Maxwell-Boltzmann rules; they obey Fermi-Dirac rules, instead. What happens to them as T increases? The states in the box are separated by finite intervals; it takes a certain amount of added energy for an electron in a filled state to rise up to a higher unfilled level. (Remember that the Fermi-Dirac statistical rules do not allow an extra electron in a state that is already filled.) If thermal energy kT is provided to the electrons, only those electrons within an energy kT of an unfilled level can make the jump—only those electrons in the upper states near the Fermi level. At low temperatures kT is small, and only a small percentage of the electrons can move up to fill a previously unoccupied level. Even at room temperature, kT is less than one percent of the Fermi energy of a typical metal.

Surely it was such reasoning that led Pauli to believe that electrons in a metal obeyed his exclusion principle, just like electrons in individual atoms. It would explain why so few electrons appeared in all the experiments designed to count them: only the electrons near the Fermi level are important, because only those electrons can easily change their energy. Pauli used this result in 1926 to explain why the alkali metals—lithium, sodium, potassium, and the like—were only weakly magnetized in a magnetic field, whereas Langevin theory predicted a spectacularly large magnetization. The magnetization was weak because only a small percentage of the electrons could be involved. The energy transmitted to the electrons by the magnetic field was too small to allow a large number of electrons to move to higher states.

Pauli had nothing but contempt for the theory of solids. Once he showed convincingly that his exclusion principle worked for electrons in metals, he lost interest in the problem and left it to his mentor, Arnold Sommerfeld, to move the pawn to the next square. At this time Sommerfeld was one of the most prestigious figures in theoretical physics. He presided over the physics institute in Munich as a debonair despot, surrounded by a small circle of outstanding young physicists, few of whom would escape a Nobel Prize. Although he was open to new ideas, Sommerfeld was nevertheless a classicist, a very clever mathematician, and a physicist with encyclopedic knowledge. An outstanding teacher, his five-volume physics course remains an excellent reference. He is known also for his semiclassical models, intuitive pictures that introduce some quantum ideas into classical descriptions and provide a basis for understanding the complete quantum mechanical theories.

This is just what he did for metals. He started with the model of Drude and Lorentz, based on billiard ball electrons, and added Fermi-Dirac statistics, the distribution rules we have just described. Thus, his electrons were classical objects, but their distribution in energy obeyed a quantum mechanical law—a good example of a semiclassical model. The number of electrons allowed to circulate in the metal under the influence of an applied electric field (electrical conductivity) or under the influence of a difference in temperature (thermal conductivity) was a small fraction of the total number of free electrons in the metal, and this fraction varied with temperature. Success was immediate and complete: all the contradictions that had bedeviled the model of Drude for the last thirty years suddenly vanished. Specific heat, the main stumbling block of the old theories, was suddenly understood. The new calculation of the Lorenz number yielded a value of 2.71×10^{-13}, in essentially perfect agreement with the experimental value.

A contradiction nevertheless lurked in this semiclassical treatment. The electrons were treated like classical particles; the only quantum mechanical aspect was the statistical distribution of their energies. To explain the large values of the conductivity of metals, however, the electrons would have to travel large distances without a collision. If we return to our pastoral analogy, it's as if the sheep were penned up in a vineyard but never encountered a vine. They could move in a straight line for a hundred times the distance between two vines, in any direction, and never trip on the smallest wandering root.

This problem seems not to have bothered Sommerfeld, who was extremely satisfied with his immense progress, but his young students found it most troublesome. They were enthusiastic about the prospects for the new quantum mechanics of Heisenberg and Schrödinger and anxious to try it out in every possible domain. Felix Bloch therefore decided to treat the electrons in metals as quantum mechanical objects, as waves of probability, following the suggestion of Louis de Broglie that had led to Schrödinger's famous equation.[3] He soon had a most surprising result: in a perfect crystal the electrons would move almost like free particles, like sheep in an open field. A completely regular and static lattice of atoms in a crystal poses no obstacle to their movement and serves only to constrain their possible speeds. It is imperfections in the crystal lattice that cause resistance. Not only did this result justify a posteriori Sommerfeld's model, but it also established a new framework with powerful tools for constructing a rigorous description of the physics of metals.

Bloch functions, the mathematical expressions for the electron waves, were developed in this pioneering work, and they continue to be used. It was the late 1920s, and the small world of physics greeted the new theory warmly. It was entirely proper at this time to speak of a circle of theoretical physicists encompassing the triangle of Munich, Zurich, and Leipzig. Deep friendships among a small number of scientists led to frequent contact and exchanges; every new idea was immediately discussed.

These new theoretical tools allowed physicists to explain various transport phenomena, including electrical and thermal conductivity, as well as magnetic, optical, and even some mechanical properties of metals. Only one domain, superconductivity, remained in the shadows, in spite of the best efforts of a glorious team, including Niels Bohr, Werner Heisenberg, Bloch, Landau, Bethe, Brillouin, Ralph Kronig, and others. The situation was so bad that Felix Bloch is said to have formulated a theorem: any theory of superconductivity must be false. A book by Nevill Mott and H. Jones that appeared in 1937 illustrates the frustration about superconductivity: *The Theory and Properties of Metals and Alloys*, as the name suggests, was supposed to be encyclopedic in its coverage of knowledge in this area. Superconductivity was not mentioned.

W. H. Keesom's handwritten notes from a meeting in Leiden on Nov. 8, 1920. Different hypotheses for understanding superconductivity were discussed during the meeting in which Ehrenfest, Einstein, Kamerlingh Onnes, Langevin, and Lorentz, among others, took part. Keesom's notes shown here describe, with the help of some drawings, a proposition by Einstein: the electrons follow spiral trajectories around each atom and then pass from one atom to the other without colliding. It is interesting to observe that this kind of "bound state" is similar to certain models recently considered for understanding high-temperature superconductivity in new materials.

Although no published paper from this period should be considered among the seminal works on superconductivity, several fundamental concepts were introduced, ideas not properly formulated that would nevertheless provide the basis of later models. The theorists of 1930 were fascinated by the ability of superconductors to support a permanent current, without any continuous external source of energy. As we have seen, the Bloch-Sommerfeld theory gives a good description of the motion of electrons in ordinary metals. It predicts that electrons cannot move in one direction indefinitely in real, never perfect, crystals because each electron undergoes collisions. Because the currents in superconductors are permanent, however, they cannot consist of ordinary Bloch electrons. Thus was conceived the idea of considering the entire collection of electrons in a metal as a coherent whole, moving together through the entire length of the superconductor. Bound together like this, perhaps the electrons would be insensitive to small, localized faults in the crystal structure, which would lead to resistance in ordinary conductors. Unfortunately, it was very difficult to imagine by what miracle the electrons would suddenly bond, just like that, at some magic temperature.

All sorts of wild ideas about the nature of superconductivity had been bandied about since its discovery in 1911. Nine years later Kamerlingh Onnes, Einstein, Ehrenfest, Lorentz, Langevin, and others met in Leiden to consider possible explanations. Some sketchy notes by W. H. Keesom give us the flavor of the discussion. Kamerlingh Onnes began the meeting by criticizing an old model of J. J. Thomson and proposed instead a rather peculiar model for the motion of the electrons. In a superconductor, he said, the electrons were lined up like billiard balls in rows in contact with one another. If a ball happened to hit another ball at one end of a row, it would knock the ball at the other end of the row out of alignment as well. He also proposed another model, one that took into account the fact that conduction electrons might spend part of their time orbiting the positive ions of the lattice. The paths of the electrons would wind through the metal like film in a movie projector. Lorentz objected immediately that this model wouldn't work for three-dimensional motion. The discussion then turned to the behavior of superconductors in magnetic fields. Einstein couldn't decide between two models, one in which each electron made several orbits around an ion before passing on to the next ion and a second in which it made only a single orbit around each ion. Apparently, he preferred the first.

During the third Solvay Congress in 1921, Kamerlingh Onnes summarized the critical issues in research on superconductivity in his communication on "Superconductors and the Rutherford-Bohr Model of the Atom."[4] He wondered how atoms could have a superconducting contact with one another and how they could form superconducting filaments capable of opening a macroscopic path that conducting electrons could follow, without transmitting energy to the neighboring atoms. Einstein suggested again in 1922 that superconducting electrons would run along some kind of molecular chain inside the metal. As time went on, innumerable discussions focused on finding a model of supercurrents compatible with the new physics. Two major camps emerged, the one defended by the old guard, led by Bohr, and the other by the young Turks of the new mechanics: Bloch, Pauli, and, on a parallel track, Landau.[5]

The model that Bohr proposed was never published—Bloch had too many strong reservations. Bohr thought of the superconducting state as a kind of electronic ice in which the electrons would no longer run about freely among the atoms. Instead, they formed a kind of crystalline lattice distinct from the atomic lattice and could move as a block through the atomic lattice without interacting with it. The frictionless motion of the electron lattice was the supercurrent. At the critical temperature the electronic ice simply melted, so that individual electrons could once more move freely. This model was simply too classical for the fiery Bloch. Through long discussions, both direct and by letter, Bloch managed to persuade Bohr to give up an idea with so little physical foundation. He was less successful with another physicist of that generation, Ralph Kronig, who had, independently of Bohr, also constructed a model of crystallized electrons, which he insisted on publishing.

In addition to superconductivity, ferromagnetism was another property of solids that demanded attention at this time. Iron, nickel, and cobalt are the classic examples of ferromagnetic materials but not the only ones. Such materials become strongly magnetized when placed in a magnetic field, and some retain this magnetization when the external magnetic field is removed. A whole series of metallic oxides behave this way; the best known are the ferrites, nonmetallic ceramics that are compounds of iron oxide and other metals. The naturally occurring mineral magnetite, which, like the word *magnetism* itself, owes its name to the ancient city of Magnesia, is a mix of iron oxides. The little magnets that attach notes to our refrigerators and keep our cabinet doors shut, magnetic tapes, and computer disks all use ferromagnetic materials.

Because electrons act as if they are spinning, and so create "spin currents," electrons themselves behave like tiny magnets of known strength. If the north poles of all the electrons in an iron bar were lined up in one direction, and thus the south poles in the opposite direction, all the ferromagnetic effects of the bar could be easily explained. Unfortunately, there was no force between the electrons known in classical physics that might align the electrons like this. Even worse, Paul Langevin had shown in 1905 that thermal agitation of these electrons hindered the tendency of their magnets to align themselves in an external magnetic field like compass needles. In addition, Pierre Curie had earlier discovered that, just like superconductors, many magnets can be magnetized only if the temperature is low enough. In the absence of a classical theory of these effects, it became urgent to find a solution in quantum mechanics.

Heisenberg found the key in 1928 simply by applying the Pauli exclusion principle to ferromagnetism. When two electrons get close to each other, the Pauli principle allows some quantum states of the pair and forbids others; which states are allowed depends on the alignment of the electron spins. The electrons act as if there is a repulsive force between them for some alignments and an attractive force for others. Associated with this apparent force is an energy, which Heisenberg called *resonance energy* and which today is called *exchange energy*. The exchange energy is much larger than any comparable energy that had been considered in classical mechanics; it was enough to keep the electron magnets aligned at room temperatures. The only mystery was why only certain materials are ferromagnetic. This problem occupied Heisenberg, Bethe, Bloch, Pauli, and a few others from 1928 to 1932, part-time, of course. After these four years the essentials of the solution were clear, and physicists were much closer to an understanding of the interactions between electrons inside a metal.

The resemblance between ferromagnetism and superconductivity was striking to Bloch and to Landau as well. Since the days of Ampère, ferromagnetism had been described in terms of persistent currents. When Kamerlingh Onnes made the first magnet based on a current circulating permanently in a superconductor, he had explicitly considered it as the macroscopic equivalent of Ampère's permanent surface currents in magnets. With the lively encouragement of Pauli, Bloch tried for several years to find a solution of the equations for electron transport in a metal that would allow permanent currents. The more he looked, the less he found, as he established a theorem stating that the most stable state of a metal always has no current at all, thus eliminating even the possibility of a superconductor. The corollary of this theorem, of course, is his second theorem, cited earlier, formulated in a moment of pique, that every theory of superconductivity must be false.

Kamerlingh Onnes's experiment is important in some accounts of superconductivity only for its practical value; it led to superconducting magnets. But, attentive as Kamerlingh Onnes was to such applications, we know that he never lost sight of the fact that his experiments were important also for theoretical physics. This attitude is not as widespread today as you might think.

Landau, for his part, didn't trouble himself with worries over whether permanent currents were theoretically possible. Instead, he took them as given and laid the foundation of what was to become one of his greatest successes, the theory of phase transitions. He started from the idea that, when a system undergoes a transition from one phase to another, one of the phases is always more ordered than the other. When water freezes, for example, the molecules, which move about quite randomly in the liquid phase, transform into a well-ordered crystalline solid, ice. In the same way, the electron magnets, which tend to line up in one direction in the ferromagnetic state, point every which way once the temperature gets above the critical value for a phase transition, and the magnetization is destroyed.

Previously, Ehrenfest had noted that phase transitions come in two varieties, which he labeled *first order* and *second order*. In first-order transitions all the constituents of a significant part of the system change their state at once. Latent heat must be added or removed to produce the change, as when ice melts or water evaporates. All the water molecules pass suddenly from ice crystals to a flowing liquid. Energy must be furnished to achieve this kind of global change.

A second-order transition, on the other hand, involves no heat, and the passage from the ordered state to the disordered state happens progressively. The closer the approach to the transition, the less ordered the system. The transition occurs when all order has disappeared. The difference in energy between the system just before the transition and the system just after the transition is infinitesimal and no (latent) heat is involved in the actual transition. Think of bowling, with many pins set up in ordered fashion, only to be knocked down and strewn about in disarray. Some may fall on the first try, more perhaps on the second. The ordered state hasn't disappeared if even one pin is still standing. On the other hand, a first-order transition is like the collapse of a house of cards, in which there are only two possibilities, the ordered state with the cards forming a house and the disordered state with the house in ruins.

Landau wanted to obtain a quantitative description of second-order phase transitions. To do this, he had to decide on some way of measuring order—he had to find an "order parameter." The order parameter by definition would be zero for a pure disordered state and unity in the perfectly ordered state. The difficulty (and, of course, Landau's brilliance) lay in the choice of order parameter. For a set of pins that a bowler is trying to topple, a clear choice of order parameter would be the ratio of the number of pins standing to the total number of pins. Landau's decision was more difficult, but, once he made it, he could express the difference in energy between two states in terms of this parameter. This difference had to become zero for second-order transitions, since they don't require any added energy. With good sense and considerable intuition Landau proposed a mathematical expression for this energy that was completely general, valid for all second-order transitions provided two other parameters were adjusted for the particular transition at hand. Once the energy was fixed this way, all the other thermodynamical quantities, such as entropy and specific heat, became calculable. This very elegant description turned out to be remarkably efficient in many different domains. It represents one of the greatest successes of so-called phenomenological methods in physics. Physicists call a theory *phenomenological* when it isn't entirely based on fundamental principles but, rather, contains ad hoc elements that are adjusted to fit the phenomena.

By analogy with ferromagnetism, where the order parameter is directly related to the magnetization (and the currents), Landau tried to describe the superconducting state by taking an order parameter directly tied to the permanent currents. This was a bad choice (one of the few times Landau's intuition failed him), even though it yielded several correct results. In any case it foreshadowed an approach that Ginzburg and Landau would fruitfully apply in 1950.

There were several other attempts to explain superconductivity during the 1930s, but none of them got very far. Bloch and Bethe offered sharp critiques of this work, which did not always please the victims of their sarcasm. Bethe suggested that one of these authors, R. Schachenmaier, knew rather little about quantum mechanics. Schachenmaier's only rejoinder was to suggest that Bethe, who was about to leave the country, was getting out to avoid retribution for the insult. This remark becomes a little more significant when we realize that Bethe was leaving to escape Nazi persecution.

The year 1933 looked like a turning point in the history of superconductivity. Solid-state researchers had made considerable progress in five years of enthusiastic application of new ideas with a spirit of friendly competition. The group of people involved was small; they knew one another well and always kept in touch. But this fruitful harvest had its limits. Years later Rudolf Peierls, a young member of the group, recalled that, once the exciting period of initial clarification had passed, enthusiasm diminished considerably when further progress seemed out of reach. The year 1933 also marked the end of the luxurious period when one could do science, travel freely, and communicate without constraints. In superconductivity itself, 1933 was the year of a fundamental discovery, the Meissner effect, which proved to be a cornerstone of all subsequent developments in superconductivity. This discovery overturned several wrongheaded ideas that had been based on the incorrect interpretation of two experiments in the 1920s.

CHAPTER 4

Experiments and
Their Interpretations

The slow pace of development of cryogenic technology during the crisis following World War I can be explained by a lack of both people and capital for research. Kamerlingh Onnes had shown how the industrial model should be applied to science, but a thriving economy was also a necessity. Experimental work on superconductivity nevertheless continued in the period up to 1933, when Bloch, Ehrenfest, Landau, and their colleagues were making significant progress in the theory. The interpretation of two of these experiments led researchers astray, however, for a number of years.

A second cryogenics laboratory did not open until 1923, in Toronto. The third began in 1925, in Berlin, and our own census indicates that Kharkov, in Ukraine, was next, in 1930. During the same period, two laboratories were begun in the United States, one in Berkeley, California, and the other at the National Bureau of Standards in Washington, D.C. Piotr Kapitza, a Soviet emigré, brought about the rebirth of cryogenics in England when he created the Royal Society Mond Laboratory in Cambridge under the sponsorship of Rutherford. Following the path of Kamerlingh Onnes, these new American and English centers focused almost exclusively on the properties of liquid helium, which was no less mysterious at the time than superconductivity. The work on helium and the work on superconductivity proceeded along quite separate paths; issues of common interest, primarily theoretical, begin to appear only after 1938.

Experimental research on superconductivity can be divided into three main categories. The first, empirical studies of superconducting materials, is almost taxonomy, an activity generally characteristic of a discipline in its youth. For superconductivity, however, this kind of research has never ceased, even at the peak of theoretical success; its continuing interest is partly attributable to its industrial applications and partly to the theoretical importance of finding patterns among the properties of new superconducting materials. Continuing discoveries showed that superconductivity is a phenomenon that is not limited to just a few metals but is, in fact, widespread.

The study of thermal properties of superconductors was the second focus of research, and it led to the first description of superconductivity based on thermodynamics. The third category consisted of investigations of the magnetic properties of superconductors. Experimental data of all sorts were produced, and it was difficult to discern the unifying elements that would lead to a coherent picture. This is an example of a problem typical in science. Two theoretical interpretations of these data have had lasting influence. One is based on the naive idea of superconductors only as perfect conductors and led to misinterpretations of some experiments. The other, which turned out to be the most fruitful, was due to London and was not readily accepted at the time. It led to the realization that superconductors could no longer be considered simply as perfect conductors. If only the physicists and chemists of 1986 had been more familiar with the history of this period when they began the race to find high-temperature superconductors. They might have published just as much, but their papers would have been more convincing.

With three different lines of research in diverse social contexts, each of the centers of cryogenics devoted to superconductivity had its own distinctive profile. Leiden, which will be our focus for the moment, investigated the general properties of superconductors. Kamerlingh Onnes died in 1926, and the Natuurkundig Laboratorium soon bore his name. The style of doing science that he developed remained alive, but, as often happens when an emperor dies, one person could no longer carry the full weight of leadership. Thus, Willem Keesom and Wander de Haas, both first-rate experimentalists, shared the management of the Kamerlingh Onnes Laboratory. Keesom became responsible for the liquefaction factory as well as research on the properties of helium and other gases. De Haas was to specialize in studies of the electrical, magnetic, and optical properties of matter at low temperature. This arrangement was accepted, but no one would say that Keesom and de Haas had a cordial working relationship. The power sharing arrangement led both sections of the laboratory to focus on superconductivity, and both made important advances in the manner peculiar to Leiden.

A young Soviet couple, Lev Shubnikov and his wife, Olga Trapeznikova, were among the numerous researchers visiting Leiden around 1930. Shubnikov set out with de Haas to measure how low-temperature electrical conductivity depends on the crystal orientation and invented a way to make the extremely pure bismuth crystals they needed. On their return to the Soviet Union, Shubnikov and Trapeznikova helped create the first low-temperature laboratory in Kharkov, where Lev Landau was still teaching (he stayed only a short time, as we shall see). Shubnikov put his crystal-making skills to good use by making samples perfectly suited for precise and surprising measurements of the magnetic properties of superconductors. His results brought him some notice, but their importance was not really understood until after World War II. He found that for numerous superconductors, particularly alloys, the magnetic susceptibility—which measures how a material behaves in an external magnetic field—varied smoothly as the strength of an external magnetic field in-

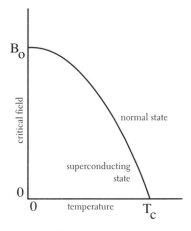

Plot of the critical field as a function of the temperature. The fact that the superconducting state represents a "phase" in the thermodynamic sense of the term did not become evident until the limits on the existence of this phase were clearly defined. The curve shown here provides the values of the critical magnetic field for temperatures below the critical temperature T_c; the superconducting state exists only for values of B and T below this curve. For example, the critical field is B_0 at $T = 0$, and it is 0 at $T = T_c$. The existence of a continuous curve that separates the normal and superconducting states and is characteristic of a particular material provided the first proof that the superconducting state is indeed a phase.

creased, while their resistivity remained rigorously zero. In other words, variations in electrical and magnetic properties no longer went together hand in hand when the superconductors were placed in a magnetic field.

Developments took quite a different turn in Berlin, where researchers systematized the search for new superconducting materials. The organizing principle in this work was a direct descendant of the German tradition in chemistry, with a decentralized network of universities in alliance with industries and their research laboratories. As a result, Walther Meissner had highly sophisticated techniques at his disposal with which to develop his materials. By tradition, the chemists at his Berlin-Dahlem laboratory studied very hard materials, such as nitrides and carbides of transition metals, used for cutting, drilling, and grinding tools. The carbide bits used for drilling in concrete are a well-known example. Transition metals — including, for example, niobium, titanium, and zirconium — can be found in the middle of the periodic table, halfway between the alkalis such as lithium and sodium at one end and the rare gases such as helium and argon at the other. Meissner's group was thus able to study transition metal compounds, which became superconducting at temperatures as high as 10K. They were the first specialists in materials research to be interested in low temperatures.

Meissner and his collaborators found about forty substances that were superconductors, each with a particular value of the critical temperature T_c, where it became superconducting, and critical magnetic field B_c, where superconductivity

TABLE 4.1
Critical Temperatures and Critical Magnetic Fields for Various Elements

ELEMENT	T_c (K)	B_c (TESLA)
Aluminum	1.19	9.9×10^{-3}
Lead	7.18	8.03×10^{-2}
Mercury	4.153	4.11×10^{-2}
Niobium	9.2	2.6×10^{-1}
Tin	3.72	3.06×10^{-2}

disappeared. (Magnetic fields in general are labeled B.) They discovered a general feature of superconductors: their critical magnetic field depends on the temperature in a way that can be closely approximated by a simple equation. We will generally speak of the "critical field" to mean the value at absolute zero; for higher temperatures this becomes "the critical field at T kelvins." Because it is impossible to reach absolute zero, the precise value of the critical field can't be directly measured, but its value can be determined with little error by extrapolating the behavior of the critical field near $T = 0$.

The Berliners also discovered that alloys of metals could become superconductors even if the metals themselves were not. Even more surprising was the observation that insulating chemical compounds could be superconductors when the constituents themselves were not. In ordinary metals, then, superconductivity did not seem to depend primarily on crystalline order or purity; the atoms themselves did not seem to be directly involved. The phenomenon appeared to depend only on the charge carriers, the electrons.

The working image of a superconductor that the physicists of these laboratories carried with them is roughly the one we have presented up to now. Even in 1925 opinion was unanimous that the only difference between a superconductor below its critical point and the same material at a higher temperature was the value of the electrical resistivity. In addition, it was known that a sufficiently strong magnetic field could destroy superconductivity. This property was also observed by measuring the resistivity; no magnetic measurements were done. Thus the basic questions: Was zero resistivity the only criterion for a "superconducting state"? Was the transition from the normal state to the conducting state really a phase transition?

With so little initial data, advances could come only from further experiments. The sophisticated apparatus deployed in these investigations revealed no change in the crystallographic structure, no discontinuity in the thermal conductivity, and no latent heat associated with the appearance of superconductivity. It wasn't easy in 1930 to do X-ray crystallography on a sample resting in liquid helium, but it was done well, nevertheless. The simultaneous absence of both latent heat and any change in structure was particularly troubling, because it is the presence of one or both that defines a phase change in the sense that water and ice represent distinct phases. Latent heat, remember, is the quantity of heat that must be supplied to change the state of one unit of mass. Thus, 333.5 kilojoules, the latent heat of ice, will melt one kilogram of ice; a refrigerator removes just that amount of heat to freeze one kilogram of water. In addition, the structure of ice is certainly different from the structure of water. Experiments indicated, however, that no such latent heat accompanies the superconducting transition in lead (in the absence of an external magnetic field), and the structure of the lead is unchanged.

The very idea that superconductivity could be an equilibrium state of matter was thus controversial until 1933. The dialogue at that time between theory and experiment became a tug of war between the idea that superconductivity must be some kind of nonequilibrium phenomenon and the conviction that it must be a new kind of phase transition. Such problems indicated that thermal measurements would be crucial. The first measurements of latent heat in the superconducting transition were carried out on lead and, as noted earlier, found none at all.[1] Specific heat measurements, carried out simultaneously, also revealed no anomalous behavior. Did this really mean that the idea of a superconducting phase had to be abandoned, or was it, in fact, just an unfortunate choice of experiment? Were there thermal changes related to the onset of superconductivity that just weren't visible in this experiment? Superconductivity at the time was in a situation, frequent in science, in which theory and experiment seemed to be sailing in different seas.

Experiments, however, don't say yes or no; they must be interpreted. A measurement can be correct but still give rise to error if it is badly interpreted. A negative experiment can never be taken as an absolute refutation. The best it can determine is that, within the precision of that particular measurement, no effects of the sort hypothesized were observed.

In this case, however, it wasn't just one experiment. In a series of articles between 1930 and 1934 Keesom (with Van den Ende for lead and zinc and then with Kok for tin) showed that the latent heat was indeed always zero. Nevertheless, thermal effects in the region of the superconducting transition temperature were not absent. Keesom and Kok in the same series of experiments found to their great surprise a discontinuity in the value of the specific heat for tin at the critical temperature. This was the first property beyond resistivity that was linked to superconductivity. The fact that the specific heat was known to be at least partly electronic in origin gave support to the idea that electrons were responsible also for superconductivity. It had taken twenty years to reach this point. Still, physicists hesitated to speak of a superconducting phase in the usual thermodynamics sense, because they had observed the new property only in tin and not, for example, in lead. In addition, they had seen no latent heat in either one.

The apparent contradiction between the specific heat measurements on tin and lead was soon resolved, and some general statements could be made. The electrical conductivity of superconductors is due only to the electrons. For thermal properties, however, the situation is more complex. When a metal is heated, the thermal energy is divided between the electrons and the ions of the crystal lattice. Thermal measurements reveal two characteristic temperatures in a superconductor: the critical temperature at which superconductivity appears and the Debye temperature, a measure of the hardness of the material. The Debye temperature can be considered as the upper bound of the low-temperature regime, where thermal agitation of the crystal lattice can be represented by the superposition of elastic vibrations with discrete frequencies whose quanta are called *phonons*. Lead has a low Debye temperature, 105K—which means that the density of phonons excited at low temperatures can be large—and a high critical temperature, 7K. Tin has a Debye temperature almost twice as high, 195K, and it becomes a superconductor only when the temperature gets down to 3.7K. At their respective critical temperatures, then, the phonons in lead, because of their density, carry much more energy than the phonons in tin. They mask the contribution of the electrons to the specific heat of lead, so any discontinuity is small. In tin, on the other hand, the electron contribution to the specific heat is large and the discontinuity is striking. Lead was simply an unfortunate choice for the first experiments on specific heat capacity.

Before the thermodynamic consequences of these results could be developed, one more obstacle to a proper understanding of the true nature of superconductivity had to be overcome, the consequence of a second unfortunate experiment. The suspect results of this experiment seemed to suggest that superconductivity was infinite conductivity and nothing more and to establish an unfounded parallel between ferromagnetism and superconductivity. The idea of frozen flux in a superconductor is at first sight comparable to the permanent magnetization of a ferromagnet (which really is frozen flux). Experiments on persistent currents that preceded this experiment are closely related.

If a superconductor is only a perfect conductor, a conductor without resistance — a phenomenon that the French call *supraconductibility* — it should obey Maxwell's equations for infinite conductivity. These equations link electric fields and magnetic fields at every point in space; they unify electricity and magnetism in a single framework. In fact, Maxwell himself had posed the question of how a conductor with infinite conductivity would behave, and he had used his equations to calculate how it would react if placed in a magnetic field.

The way to test whether Maxwell's equations were applicable to superconductivity was straightforward; physicists of the period who had all at one time or another made a motor or a generator readily understood the procedure. Put a ring of superconducting wire in a magnetic field. The size of the wire isn't important if it has no resistance. The plane of the ring should be perpendicular to the magnetic field lines, the imaginary lines that trace the direction of the field, so the field lines cut through the ring like a lion jumping through a hoop. We say that the magnetic flux is large. (If the plane of the ring is roughly parallel to the field lines, the magnetic flux is small.) The recipe for the experiment is then to cool the ring down below the critical temperature *while it is in the magnetic field* and then cut off the magnetic field. The changing magnetic field should induce a current in the ring, just as it does in an ordinary electric generator, with one difference: with no resistance the current should last indefinitely. It should be a persistent current, and the magnetic field created by this current should be *just as intense* as the original magnetic field that was cut off. The original magnetic field should be "frozen in" the area surrounded by the ring.

By the same token, if the ring is cooled down when it is not in an external field, the frozen flux will be very small, just that due to the earth's field. Should a magnet then be turned on, a current will be induced in the ring, causing a magnetic field of its own that exactly cancels the external magnetic field inside the ring. In both situations, cooling down before or after the external magnetic field is turned on, the field in the area surrounded by the ring is "frozen." The final field is the same as the original field; the ring "has a memory."

This experiment, which was purely conceptual for Maxwell, was carried out in Leiden by Kamerlingh Onnes (described in chap. 2). The idea was taken up again several years later by Josina Jonker and Hendrik Casimir, a pair of young physicists who later married. (We met Casimir previously through his memoirs.) They had a setup that had a superconducting ring cooled down below its critical point while in an external magnetic field. Early one evening they disconnected the batteries that drove the current in the magnet. Then, all night long, they monitored the magnetic field in the ring with a very sensitive magnetometer. (They took the precaution of doing their experiment at night to reduce background disturbances from, for example, people turning electrical equipment on and off.) They observed no change in the strength of the field, just as Maxwell had predicted for a ring with infinite conductivity.

The experiment was repeated many times, and not merely in Leiden. The most spectacular presentation, with really striking results, was the work of J. C. McLennan in London. He had apparently set out to renew with liquid helium the series of public performances that Dewar had begun. In 1934 McLennan wrote:

> Few Londoners have had the occasion to see this rare liquid that is so difficult to obtain, except those who have heard my lectures in June, 1932, and July, 1933 . . .
> The lifetime of the persistent currents seemed to be limited only by the amount of refrigerant, the liquid helium, available. During my lecture on June 3, more than two years ago, at the Royal Institution, I exhibited a lead ring immersed in liquid helium carrying a current of 200 amperes. The current had been started in the ring six hours earlier, in the afternoon, by Professor Keesom in Leiden. The current persisted without attenuation while it was transported in the liquid helium from Leiden to London by Colonel Master de Sempill (now Lord Sempill).

After World War II, this fascinating experiment on permanent currents in superconducting rings was repeated, this time with much more sensitive instrumentation but no young lovers. If the lovers had been there, they would have had plenty of time to quarrel, because the ring was left in liquid helium for two years. The cryostat (the helium dewar), of course, had to be refilled whenever the liquid helium started to run low. No attenuation whatever in the current was observed,[2] to one part in 10^{23}. The resistance really seemed to be zero. Indeed, the resistance of a superconducting ring is consistent with zero to an accuracy that is just about as good as can be measured for any quantity.

> Before proceeding, a little digression on measuring zero might be interesting. There are few physical quantities that one can reliably say are closer to zero than 10^{-23} (i.e., 1/100,000,000,000,000,000,000,000). The probability that a proton, the nucleus of the hydrogen atom, might decay into other particles has been measured to be zero with better precision. Particle physicists have been led to consider whether the lifetime of the proton is shorter than 10^{32} years. Cosmologists tell us that such an instability, if we can call it that, would have important consequences for their conception of the universe. In caves and tunnels far underground, where the earth itself acts as a shield against background particles coming from the sky, particle physicists have spent years now looking for recognizable traces of the disintegration of a proton. Their apparatus weighs thousands of tons (to be compared with the several tens of kilograms for a very large helium cryostat). The decay probability is zero, according to their measurements, to within 10^{-30}.

Persistent currents with their frozen flux in superconducting rings seemed to be well understood in terms of Maxwell's classical equations with infinite conductivity. Thus, there was no reason to suspect that a superconductor was anything other than a perfect conductor. This conviction was so well established in the minds of physicists that it led Kamerlingh Onnes himself to a hasty interpretation, later shown to be false, of a similar experiment that had been carried out in Leiden in 1924, this time with a hollow lead superconducting sphere instead of a superconducting ring.[3] The proper interpretation of the experiment with hollow spheres is subtly, but crucially, different from the experiment with rings. To complicate the matter even further, the data from the hollow sphere experiment proved to be slightly incorrect.

The hollow lead sphere was plunged into liquid helium in the presence of a magnetic field. The sphere became superconducting, and, what is most important for this episode, the measurements indicated that the distribution of the magnetic field *inside* the sphere, that is, in the hollow, seemed not to have changed during the superconducting transition. Then the external magnetic field was cut off, just as in the ring experiment, and again the field inside the sphere seemed to stay exactly the same. In other words, during the whole process, it appeared that the magnetic field remained constant, just as it did in the experiment with a simple ring. By cooling the sphere in the presence of the external magnetic field, they said at the time that it looked as if they had "frozen" the magnetic field inside the sphere. Because of the previous experience with the ring, the interpretation of this experiment seemed clear: one could draw rings around the sphere, and the field inside every ring stayed the same. The result did not seem surprising, but it was wrong.

It's this story of frozen flux that seemed to define the recipe for the experiments. If the field was applied after the sample was cooled — that is, contrary to the order in the recipe — the results were different. This outcome seemed normal, since the frozen field in that case would be only the small magnetic field of the earth itself, not the intense applied external field. As Fritz London said in his Paris lectures in 1935, once the fog had disappeared:

> Some important experiments, particularly the famous experiment of Kamerlingh Onnes and Tuyn on the persistence of supercurrents in a sphere, seemed at that time to confirm a misconception. The state of a superconductor did not seem to be uniquely defined by the usual variables [e.g., temperature, strength of the magnetic field, etc.] but rather to be characterized by some kind of memory of how it became superconducting [e.g., whether it became superconducting before or after the external field was applied]. Following this idea, the present state of a superconductor would depend on the path it followed during its history.[4]

This misconception would mean that the superconducting transition could not be a phase transition in the usual thermodynamic sense. At a given temperature and pressure, ice is ice; it doesn't matter how long it took to reach that temperature, for example, or whether it was cooled before or after the pressure was increased. Phase transitions are reversible, but the superconducting transition seemed to be irreversible, because of the apparent frozen flux. The idea of frozen flux inside the actual superconducting metal was, however, only a myth, but one with a hard skin. It depended on measurements of the magnetic field inside the lead sphere, and these measurements were not easy. The method, which had been classic since the beginning of the century, was to put a small piece of bismuth wire at appropriate positions in the sphere. The resistivity of bismuth varies strongly in the presence of a magnetic field, and the variation is particularly large at low temperatures. For once in Leiden, the credo of Kamerlingh Onnes was obviously not followed. After the frozen flux had apparently been discovered, no one took the trouble to repeat the measurements. Physicists spoke as if frozen flux and persistent currents were part and parcel of the same phenomenon. If researchers had taken the trouble to do really precise measurements, however, they would certainly have observed that the field inside the sphere is always less than the initial field. Then they would have been forced to admit that the extrapolation from the ring (where the field is indeed frozen in) to the sphere was not valid.

The results and interpretation of this erroneous experiment were diffused widely and supported by Leiden's reputation for rigor. They had a considerable, durable influence on experimenters' conception of superconductivity. The fact that the superconducting state seemed to depend on the history of the material indicated that the superconducting transition was not an ordinary thermodynamic phase transition. This explains the persistent reluctance of the experimenters to adopt the point of view that there was a phase transition. It is thus all the more remarkable that theorists such as Landau and Bloch had few qualms about believing in a superconducting phase right from the start, even though they didn't know how to describe it.

When Fritz London published his version of the electromagnetism of superconductors (in French, for reasons we'll return to), he introduced his subject with the following attempt to draw an analogy between ferromagnetism and superconductivity:

> Bloch and Landau formulated a program whose realization has generally been considered as the task of a future theory of superconductivity. It seemed necessary to imagine a mechanism that, without any external field, would make it possible for a metal in its most stable state to support a *current*. The thermodynamic stability of the superconducting state and in particular the stability of the persistent currents themselves seem necessarily to lead to this idea. In this connection, one often thinks of the example of ferromagnetism, where the most stable states consist of permanent magnetization without the involvement of any external field.[5]

The myth of frozen flux, which would ultimately disappear, was starting to fade away. It took a long time for London's point of view to be accepted. The fact that he was working in exile did not help.

CHAPTER 5

The True Image
of Superconductivity

The two misleading experiments that impeded progress in understanding super-conductivity during the 1920s were reinterpreted at the beginning of the 1930s. This activity finally led to the discovery of the one property truly characteristic of a superconductor, not zero resistance but perfect diamagnetism: in the presence of an external magnetic field, a magnetic field is induced in the superconductor that exactly cancels the external field. The net magnetic field inside a superconductor is then zero, just like the electric field. This effect, discovered by Meissner, does not necessarily follow from Maxwell's equations for a perfect conductor, but it is not excluded by them either. In fact, it turns out that superconducting materials behave quite differently as their temperature is lowered below the transition temperature if they are in an external magnetic field. In the language of Ehrenfest the transition will then be first order and thus involve latent heat. Without an external field a second-order transition will occur, without latent heat, but this distinction was not clear in the early 1930s.

Once again the Dutch made essential contributions.[1] De Haas and H. Bremmer found in 1931 that there was no discontinuity in the thermal conductivity at the critical temperature of tin in the absence of a magnetic field but a marked discontinuity if the superconducting transition took place in a magnetic field. To the experimenters it looked like more bizarre behavior from superconductors. The experimenters were looking for an indication of latent heat, the traditional signature of a phase transition, and found something of interest only when the sample was in a magnetic field. These thermal experiments in a magnetic field might have argued in favor of a phase change, but the discontinuity was instead attributed to the presence of trapped, "frozen," flux.

Faced with these apparent contradictions, some experimenters began to think once more about the famous experiment of Kamerlingh Onnes and W. Tuyn with the lead sphere. Perhaps they had accepted the facile interpretation of frozen flux too readily. W. De Haas, J. Voogd, and J. Jonker therefore tried a number of experiments on different kinds of tin wires. When they used monocrystalline wires they

Waltter Meissner in his laboratory in Berlin.

soon observed that the order of cooling and applying the magnetic field made no difference whatsoever to the field measured inside the wires. The order could be reversed, and nothing would change; the monocrystalline superconductors did not remember their history. This was hard to reconcile with the notion of frozen flux, but it was suggestive of a phase transition.

Ehrenfest introduced the notion of second-order phase transitions in 1933; his description was supposed to be quite general, valid for all second-order phase transitions. In fact, second-order phase transitions had already been observed by this time, in ferromagnets at the Curie point, and they had been analyzed by Edmond Bauer in 1929.[2] A. J. Rutgers, Ehrenfest's graduate student, thus decided to apply his description to superconductors, with no justification other than simple intuition. He looked for a relationship between the shape of the critical magnetic field curve and the discontinuity in the specific heat at the critical temperature. Even though these two quantities had seemed to be totally independent of each other, Rutgers's equation was in excellent accord with the experimental data, an indication that the superconducting transition without an external magnetic field is a second-order phase transition.

If the superconducting state did indeed correspond to a phase distinct from the normal state, something Bloch and Landau had never doubted, thermodynamic reasoning could then be used to describe it. C. J. Gorter, a young experimenter from Leiden, decided to concentrate his efforts in this direction, and he linked up with Casimir, another of Ehrenfest's students, who had been closely following the work of de Haas on superconductivity. Casimir, who was going to marry Josina Jonker in August of that year, 1933, had just returned from a year with Pauli in Zurich and was losing confidence in the idea of frozen flux.

C. J. Gorter in Leiden.

Gorter and Casimir started to work together, while Jonker thought about redoing the experiment on the lead sphere. She was not alone. Word soon came from Berlin that W. Meissner and R. Ochsenfeld had demonstrated that a superconductor expels magnetic flux from its interior.[3] The Berlin group had been carrying out measurements of the magnetic field produced outside tin and lead samples when they discovered this phenomenon that was exactly the opposite of the expected frozen flux. Not only was the magnetic field not frozen inside the sample; it was actually chased out—the magnetic field inside was zero ($B = 0$). Meissner's experiment started a wave of short publications from cryogenics labs around the world. Gorter reacted immediately and sent a note to *Nature* with the seemingly brash suggestion that $B = 0$ is a general characteristic of all superconductors. The myth of frozen flux was shaken, even if the evidence from different experiments was not yet entirely consistent.

Meissner's experiment provided an immediate justification for the theory that Gorter and Casimir were constructing, and they soon published what would be the first thermodynamic approach to superconductivity. Their treatment, the *"two fluid" model*, proved exceedingly popular in the small world of low-temperature physics, and it would remain so for many years, even during the 1950s, when new experiments would show how rudimentary it really was. When their article appeared, however, it was the thermodynamic approach that captured people's attention.[4] Gorter and Casimir considered the superconducting state as a phase, more ordered than the normal phase. They suggested that a magnetic field can penetrate into a super-conductor in the normal phase, while it is excluded in the superconducting phase. At the transition temperature the metal was equally likely to be in one phase or the other, and so the "entropy" (related to the entropy in thermodynamics, a measure of disorder) had to be equal in the two phases at T_c. They demonstrated that the formula that Rutgers had intuitively proposed had, in fact, a solid foundation in thermodynamics.

In September 1934, in spite of the political situation in Germany at the time, Gorter went off to a meeting of the German Physical Society devoted primarily to low-temperature physics. Thirty years later he justified his attending this meeting at Bad Pyrmont by suggesting that "at that time only a few German physicists had been influenced by the Nazi ideology."[5] According to Gorter, Meissner left to him the responsibility of defending the proposition that the magnetic field was strictly zero inside a pure superconductor. In fact, many experimenters were slow to accept the Gorter-Casimir theory wholeheartedly. Thermodynamic arguments were quite familiar to most low-temperature physicists; they were much less specialized than today's physicists and were used to dealing every day with thermodynamic quantities. Nevertheless, they were skeptical about the proposition that the magnetic field was always zero inside a superconductor, and even Meissner didn't endorse it right away.

The experimenters were still cautious three years later, in 1937; maybe they were simply overwhelmed by an avalanche of theoretical papers. In any case we can hear their skepticism in the remark of an English physicist working in Kharkov with Shubnikov: "Gorter considers the supraconducting state as a phase in the usual sense."[6] Nevertheless, the Gorter-Casimir theory turned out to be popular for a long time because of its straightforward description of the electrons. At the superconducting transition point, it suggested, the electrons divide into two fluids, one of which feels no resistance. The percentage of superconducting fluid is unity at absolute zero and zero at the critical temperature. Finally, we have a simple — too simple — representation of superconductivity. The Gorter-Casimir model accepts the complete expulsion of magnetic field from the superconductor, but it says nothing about how the circulation of two fluids makes it come about. By writing $B = 0$, however, the two men made a definitive break with the idea of a superconductor as simply a perfect conductor. Meissner and Ochsenfeld had done this with their experiment; Gorter and Casimir did it in their theory, without saying a word about electrodynamics or paying any attention to quantum mechanics.

Fritz London, another theorist, attacked the same problem. He was already in exile from Germany and didn't have to ask himself whether a majority of German physicists were Nazis or not. Right after Meissner's publication, he and his brother Heinz began a calculation similar to Gorter and Casimir's before making the annoying discovery that the latter had already been published. It was only later that Fritz took up the problem of the electrodynamics of superconductors. In his lectures of April 1935 at the Institut Henri Poincaré in Paris he says: "Our ideas have recently undergone a profound revolution brought about by an experiment first carried out by Meissner and Ochsenfeld in 1933 and repeated since then by several others."[7] The others included Y. N. Riabinin and L. V. Shubnikov,[8] as well as K. Mendelssohn and J. D. Babbitt.[9] What these experiments revealed is that the Meissner effect indeed exists, but it can be properly observed only in samples that are monocrystalline and approach physical perfection. M. Ruhemann, who closely followed the work of Shubnikov and Landau at Kharkov, had this to say:

> It appears quit clear that, if the condition B = 0 were really a characteristic property of the supra-conducting state, we should soon have a satisfactory picture of the phenomenon of supra-conductivity. Nevertheless, experiment is at present working in another direction, and the exceptions to this rule are now more the centre of interest than the rule itself. . . . A subsequent full theory of supra-conductivity will thus have to account for two phenomena, the disappearance of resistance and the behaviour of the magnetic flux. Hitherto it is not yet apparent which of the two is the essential and primary effect, supra-conductivity or "subpermeability."

Ruhemann wrote these skeptical words after London's theory appeared; the last lines served as the conclusion of his text on superconductivity published in 1937. Fritz London himself had to take into account the same concerns, as he said in his lectures:

> The fact that magnetic flux has been found frozen inside superconductors under "non-ideal" conditions seems linked to the presence of non-superconducting inclusions. This might be explained by the inhomogeneous nature of the material. On the other hand, maybe there are magnetic lines of force in certain regions of the superconductor that are so dense that the field there is above the critical field, and the superconductivity is destroyed there. The presence of frozen fields would then be uniquely tied to the presence of *superconducting rings* enclosing the magnetic flux in the non-superconducting inclusions.
>
> As a result, we have to regard the persistence of the initial magnetic flux not as a fundamental phenomenon but rather as the complex result of the fact that there are several constituents or phases superimposed on one another in a multiply connected topology. On the other hand, the elementary phenomenon of a *pure* superconducting phase would be something much simpler. Given Meissner's discovery, it appears that the transition from the normal state to the superconducting state is *reversible*, in the sense that the magnetic field *is always equal to zero* at every point of the superconductor, *independently of the path followed*, whether the external magnetic field was applied before or after the sample is cooled through the superconducting transition.

For London, then, a superconductor was characterized not only by the disappearance of the electric field E (E = 0 inside a superconductor) but also by the disappearance of the magnetic field B (B = 0 inside the superconductor).

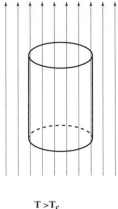

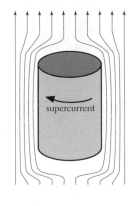

T > T_c   T < T_c

The Meissner effect, shown schematically here, is proof that the superconducting state is not uniquely characterized by zero resistance. The figure on the left shows magnetic field lines (represented by arrows) traversing a metallic sample when the temperature is above the critical temperature and the sample is in the normal state. If the sample is then cooled down below T_c while it remains in the external magnetic field, a supercurrent is generated at the surface of the sample, and the field lines are expelled. The difficulty of such experiments coupled with the unexpected result provide some explanation of why it took twenty-five years of similar experiments to discover the Meissner effect.

Everyone agreed that the electric field must be zero. Without resistance there's no possibility for an electric field or a voltage difference to exist between two points of a superconductor. In other words, a superconductor is an absolute short-circuit throughout its volume. The second condition arises from Meissner's experiment in the ideal case. To pursue his theory London went back to Maxwell's equations. In keeping with Bloch's first theorem, he supposed that there's no current in a superconductor when it is not in a magnetic field. When it is in a magnetic field, there must be a current of superconducting electrons on the surface that creates a magnetic field that opposes the external field. As London explained: "For a superconductor, the magnetic field is not simply a natural accompaniment of a current, as it is for a normal conductor. It plays another role, an active role, so to say. We consider it to be the *cause* of the current in the superconductor, just as an electric field is the cause of a current in an ordinary conductor." To produce a current in an ordinary conductor, an electric field must be applied, and the current is proportional to the electric field. It's the electric field that gives rise to the current. In the language we have used previously, a voltage difference across a wire causes an electric field in the wire that exerts a force on the electrons.

As London explained things for a superconductor, a thin layer of superconducting current, the "supercurrent," circulating around the surface induces a magnetic field that cancels out the magnetic field inside the superconductor. We have seen previously that Maxwell's equations are not sufficient to explain such a magnetic "shield" of surface currents; they suggest frozen flux, not the expulsion of flux.

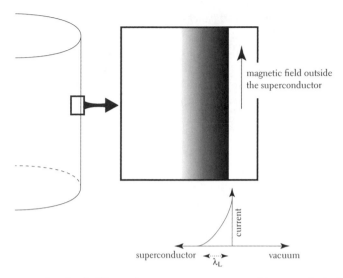

magnetic field outside
the superconductor

current

superconductor $\xleftarrow{\quad}\xrightarrow{}$ vacuum
λ_L

The London penetration depth. The surface supercurrent generated in a sample during Meissner effect experiments generates its own magnetic field. Inside the sample this induced magnetic field is equal in magnitude and opposite in direction to the external magnetic field. The absence of magnetic field inside the superconducting sample is due to the cancellation of the two fields. A model developed by Fritz London explains the existence of the supercurrents, which decreases rapidly with distance from the surface. They are thus confined to a thin shell whose thickness, labeled λ_L, the London penetration depth, is about a nanometer (10^{-9} m) and thus difficult to measure.

Thus, Fritz and Heinz London added a new equation (the London equation) that, when combined with Maxwell's equations, yielded solutions that fully accounted for the Meissner effect. There is no simple theoretical justification or intuitive explanation of the London equation. When it appeared in 1935, its justification was purely phenomenological; perhaps this explains why once again many physicists were slow to accept it. It is easy, however, to describe the form of the solutions to the equation.

Take, for example, a cylindrical sample of tin and put it in a magnetic field that's not strong enough to destroy superconductivity, maybe a hundred times the earth's magnetic field. Place it so that the axis of the cylinder is along the direction of the magnetic field. As long as its temperature is above T_c, the tin is in its normal state, and the magnetic field penetrates the volume of the cylinder almost as if the cylinder weren't there. Below the critical temperature for that magnetic field, the cylinder becomes superconducting, and the magnetic field lines will now bypass the cylinder, since they can no longer get inside.

A supercurrent will then circulate around the cylinder in a thin layer and create a magnetic field that exactly cancels the external field inside the cylinder. The thickness of the supercurrent layer can be calculated with the London equation. Within the layer there is in fact a magnetic field; right at the surface of the superconductor its strength is the same as the strength of the external field. The thickness of the layer is called the *London penetration depth*, traditionally represented by the notation λ_L. It depends on the metal and on the temperature; well below the transition temperature, it is typically about 10^{-5} cm. Maxwell's equations, together with the London equation, show that the magnetic field cannot penetrate beyond this surface layer. A good test of the London equation, then, is to measure how far a magnetic field penetrates into a superconductor. Such an experiment was not carried out until 1940, when Shoenberg measured the penetration depth in tiny spherical beads of mercury.

The work of the London brothers put a stop to all the aimless wandering in research on superconductivity. Gorter and Casimir had demonstrated that one had every right to consider superconductivity as a phase in thermodynamic equilibrium. The transition to the superconducting state in a magnetic field was reversible; the sample did not remember its history. Beyond this the London brothers showed that the Meissner effect should not be interpreted as an indication that the permeability of superconductors is zero.

> The magnetic permeability of a material describes how a material adapts to an external magnetic field. A ferromagnet is an extreme example. The permeability of a ferromagnet such as iron might be thousands of times the permeability of a nonmagnetic material, and the material might remain magnetized even in the absence of the external field. What has happened is that the external field has aligned many of the tiny electron magnets, and they stay lined up after the external field has been removed.
>
> Ferromagnetism is a strictly local property of the material. Spins are aligned in microscopic regions called domains, and there is no observable difference between what happens to the interior and the surface of a magnetized bar. The total absence of a magnetic field in a superconductor is, on the contrary, a global phenomenon that results from the circulation of macroscopic currents. It is this macroscopic character that served as a starting point for the theoretical development that played a decisive role in understanding superconductivity and superfluidity. It was also the key to applications of superconductivity to scientific instrumentation.

The theories of Gorter and Casimir and of London were received quite differently. A review article entitled "Superconductivity and the Theory of Metals," written in 1936 by A. H. Wilson, who had just completed an influential book on the theory of electrons in metals, reveals this discrepancy quite clearly. The Meissner effect was decisive for him in deciding that the superconducting state was indeed a phase in the thermodynamic sense, and he went on at length about it. London's theory, however, merited but a brief summary. Wilson gave him credit for proposing "a most promising angle of attack" but found it "purely formal." [10]

The scope of London's work was not immediately appreciated. Nevertheless, it was this work that marked the beginning of the modern approach to superconductivity. Two different developments help explain why it was not an instant success. Shubnikov's experiments, for one, showed that the Meissner effect was only imperfectly realized in many situations. Landau interpreted these results as evidence for a state intermediate between the normal and the superconducting states. Today we know from the work of Landau himself that this idea is applicable to only a certain type of superconductor. The second development was a new qualitative model of the behavior of superconductors in a magnetic field proposed by Mendelssohn, then a German exile in England. He tried to explain the numerous situations in which the Meissner effect was not complete by comparing the penetration of magnetic field into a superconductor with the penetration of water into a sponge. Mendelssohn is the author of an excellent book on the history of low-temperature science, which we have cited here many times, as well as numerous important articles in the field.[11] Ironically, however, he is best known for a model that turned out to be false, whose great success was based on an image that was easily comprehensible but had no predictive power.

During this period, however, London did find one attentive reader: John C. Slater, a Harvard physicist at the forefront of developments in quantum mechanics and atomic physics. Slater discussed an article of the London brothers with a student of his, John Bardeen, about whom we will hear more later.[12] He then wrote two articles that showed that the work of the London brothers had an important consequence:[13] the superconducting electrons had to move in large orbits—they were highly "delocalized." The Heisenberg uncertainty relations thus required that their velocities be well defined and close to the Fermi velocity. Their energies thus had to be close to the Fermi energy, much closer than the theory of metals would predict. Bardeen never forgot Slater's remarks, and he recognized that all his work on superconductivity was initially inspired by the work of Fritz London.

The authors of this book have been muddling about in the field of superconductivity for many years, and they have certain prejudices about the relative importance of each of the principal protagonists. It is Fritz London, so little appreciated in his time, who stands out to us in this large and imposing crowd. One important reason is that, toward the end of his life, he saw the evolution of physics toward big science, and, once more alone among his peers, he expressed his concern about it.

CHAPTER 6

Fritz

Today the name Fritz London evokes only a few famous results and an eponymous prize for low-temperature physics funded by John Bardeen's second Nobel Prize, this time for superconductivity. Until the biography by Kostas Gavroglu published in 1995,[1] there was only one memorable photograph of London, in front of a black-board covered with equations, his face distorted and no hint of a smile. Not much for one of the most engaging intellectuals in the pantheon of low-temperature physics. London stayed out of the everyday scientific brouhaha. Perhaps it was his lofty point of view on physics, more philosophical and conceptual than others', that made it difficult for him to spread his ideas and served to make him something of an outsider. It is thus interesting to consider the issues that he found so absorbing as well as his style of work.

What we find striking today about London's papers is his persistent desire to place each one within a general intellectual framework. There was no shortage at the time of talented physicists whose research may have been as important as London's, but few of them expressed an equal determination to achieve a coherent body of work. For example, in the conclusion of each of his articles London announced what he wanted to do next, and then he did it. In addition, the general perspectives that he established have turned out to be remarkably prescient about issues that proved important in domains well beyond low-temperature physics.

Fritz London was born in 1900 in Breslau, Germany, now Wroclaw in Poland, seven years before his brother Heinz. Their father became professor of mathematics in Bonn in 1904 and died very young, in 1917. Right after the war Fritz studied physics, mathematics, and philosophy. First came logic, and, at twenty-one, he wrote a paper ostensibly for his own use that was immediately accepted as a thesis by his philosophy professor, defended in Munich, and then published in Edmund Husserl's journal, the *Jahrbuch für Philosophie und Phänomenologishe Forschung*, in 1923. The subject was "the formal conditions of purely theoretical perception," so it was not by chance that Husserl was interested. As surprising as it seems, Fritz's itinerary was far from unique at the time. Think of Robert Musil, who wrote *Der Mann ohne*

Eigenschaften (*The Man without Qualities*) or Alexandre Koyré in Göttingen, who was studying philosophy with Husserl at the same time as he was studying mathematics with David Hilbert.

London worked for a time under the direction of Sommerfeld after taking Sommerfeld's course in quantum mechanics. In 1925 he became the assistant of Paul Peter Ewald, first at Stuttgart, then at Göttingen. It was during this period that he carried out the work for which he is still most famous outside the circle of low-temperature physicists. Collaborating with Walter Heitler during a stay in Zurich, London calculated the binding between the two hydrogen atoms in a hydrogen molecule, the first quantum mechanical approach to the chemical bond.[2] This work marked the birth of quantum chemistry and had a lasting influence on London's style and intellectual preoccupations. Nevertheless, it seemed to lead nowhere at the time. The theory of the chemical bond in other systems demanded complicated calculations that could not be avoided except by simplifying assumptions. Thus, London set out to understand the dynamics of the interactions between atoms and between molecules, including the Van der Waals interaction, which has a much longer range than typical interatomic spacings.

London's old friend L. W. Nordheim reports that it was around this time that London started to think about the possibility of transporting information from one end of a long chain of molecules to the other.[3] The molecules would then be sensitive to the macroscopic geometry of the sample. In other words, London speculated very early about applying quantum mechanics to molecules in order to consider possible macroscopic effects. At a time when quantum mechanics was successfully applied only to microscopic phenomena, this ambition must have seemed surprising. In fact, London was practically alone among physicists in attempting to probe deeply in this direction. Certainly, he was understood when he spoke of "molecular physics," but few found it important. Interesting problems abounded in other domains, and quantum mechanics was astonishingly successful at the atomic scale. Why stray from the golden path? Even after he became an assistant to Schrödinger at the University of Berlin, London remained out of the mainstream.

A few years later, when many of the most exciting problems in quantum mechanics had been solved, the idea of single-minded attention to what we would today call condensed-matter physics seemed more natural. Most of the major players in this intellectual adventure had been forced into exile by the Nazis and had to find some new areas of specialization — not everyone could find a job in nuclear physics. It was only much later, however, that London explicitly mentioned in a publication his brilliant intuition about quantum phenomena at a macroscopic scale. Nevertheless, throughout the 1930s this intuition was already the dominant motivation for his choice of research subjects. In fact, even in his first work on superconductivity, London returned to what he called *molecular theory*. He referred to it during his Paris lectures on superconductivity, so Nordheim's comment is not very surprising. What is surprising is, rather, the fact that physicists seem to think that it was not until 1948 that the idea of macroscopic quantum order could be found in a scientific journal.

All the leading physicists of the 1930s knew one another, so they could all speak about Bloch's theorem, which had never been published. Yet it is clear that Fritz London had some difficulty in delivering his message to the world of physicists. One can only wonder how much this was due to his image at the beginning of the 1930s as a theoretical chemist. By the same token, was not the label "molecular" looked upon as pejorative by a great many of his colleagues, especially experimentalists, in an era when microphysics was king?

In any case, this initial interest in possible macroscopic quantum mechanical effects led Fritz London to look closely at what his younger brother Heinz was doing. Heinz had started physics without stopping for a look at philosophy, and he began a thesis in Breslau under the direction of an excellent cryogenicist, Franz Simon, who was setting up a low temperature laboratory.[4] For his thesis subject Heinz followed a suggestion of Walter Schottky, the inventor of the semiconductor diode, a basic element of all electronics; Schottky was working in the Siemens laboratories at the time. Heinz studied how the electrical conductivity of a superconductor depends on the frequency of the current.

Up to now we've spoken of zero resistivity only for direct currents, the sort you get with a battery and a wire, for example. The magnitude of a DC current is constant in time, and it doesn't change direction; in other words, its oscillation frequency is zero. The electric and magnetic fields associated with the current are constant in time. Heinz London wanted to find out what happened to the resistivity of a superconductor with frequencies different from zero. For a normal conductor, a little mathematical gymnastics with Maxwell's equations shows that the resistivity tends to increase when the frequency increases because high-frequency alternating current exists only in a thin layer near the surface. This is called the *skin effect* — electromagnetic waves are attenuated inside a metal. The thickness of the skin varies inversely with the square root of the frequency. Thus, if the frequency of the current is quadrupled, the resistivity will double, because the skin will be only half as thick. Would a superconductor behave like this?

Schottky's suggestion to study this problem was particularly apropos in 1933, both experimentally, because of Meissner's results, and theoretically, because of an influential article by R. Becker, G. Heller, and G. Sauter.[5] Published before Meissner and Ochsenfeld's article, it contained a complete treatment of Maxwell's equations for a perfect conductor, including the effects of the inertia of the moving electrons. This article introduced the idea that the magnetic field could vary with time only near the surface of the conductor, something like the skin effect. Here, however, in the perfect normal conductor, the current could exist only in a skin about 10^{-6} cm thick, and this thickness, much smaller than typical skin thicknesses, was supposed to be independent of the frequency. This meant that the magnetic field was constant in time throughout the interior of the perfect conductor. This contradicted neither

The London brothers, Heinz (*left*) and Fritz.

the idea of frozen flux nor the experiment of Meissner, but it did indicate that a superconductor was not a perfect conductor. Moreover, the article indicated that Maxwell's equations could give rise to characteristic lengths that were frequency independent.[6] The epistemological position of this paper is thus quite remarkable: it didn't really overturn a dominant false idea, the trapped flux, but, in fact, it revealed a new way to analyze the situation starting from a well-established theory, Maxwell's equations.

Heinz London's experimental attempts to measure the attenuation of the current at high frequency were ideal experiments to test the Becker, Heller, and Sauter analysis. If the frequency increases, the skin thickness should shrink, and it should be possible to measure a thickness as small as 10^{-6} cm. The appearance of the Gorter-Casimir paper, with its two fluids of electrons, only increased the interest in experiments with alternating currents. With a continuous current, a measurement of the resistivity of a superconductor provided little information because the superconducting fluid short-circuits the normal fluid. With an alternating current, the energy losses of both fluids could be measured at the same time. By varying the temperature, one could compare the losses due to normal electrons with those of the superconducting electrons. Heinz London had a gold-plated thesis topic, which was, no doubt, the only thing at the time that could make him smile.

Hitler had been reich chancellor since January 30, 1933. The Reichstag was set afire on February 25, probably by the storm troopers; a national boycott of Jews was proclaimed for March 31. On April 1 the Prussian Academy issued a declaration, in spite of Nobel Prizewinner Max von Laue's protests, expressing their satisfaction that they were rid of Einstein. He had announced his resignation from the academy upon his arrival in New York and had decided never again to set foot in Germany. On April 7 the Reichstag adopted a law nominally aimed at restructuring the civil service that in fact had the effect of eliminating Jews and the left-wing opposition from all government positions, including teaching positions in the universities and technical schools. During the first year seventeen hundred people were affected by these measures, which aroused little protest in academic circles. Otto Hahn tried in vain to convince Max Planck to support a petition condemning the dismissals that found only twelve signers.[7] Leo Szilard, who would later play a crucial role in initiating the Manhattan Project to build an atomic bomb, arrived in Vienna at the end of April. There he met Lord William Henry Beveridge, the theoretician of the British welfare state, and convinced him that he had to provide a welcome in the United Kingdom for scientists fired from their jobs in Germany.

In January 1934 Heinz London made a last minute defense of his thesis, one of the last by a Jew in Germany before the war. He published it only after the article by Gorter and Casimir appeared. In the meantime the political situation was getting worse by the day. The English were organizing a welcome for future refugees; public funding was requested, and F. A. Lindemann, a professor at the Clarendon Laboratory in Oxford, had covered much of Germany in a search for recruits.[8] In his travels he met Schrödinger, who let him know straight on how much he disapproved of Hitler's policies. The conversation turned to the case of Fritz London, who had a junior position at the university. At that moment it wasn't yet clear whether the racist laws would be applied to such temporary positions, but Lindemann mentioned that he had offered London a position in England. "Naturally, he accepted?" Schrödinger asked; he was dumbfounded by the answer. Lindemann told him that London had requested some time to consider the offer. "That I just cannot understand," said Schrödinger. "Offer it to me, if he does not go, I'll take his place." Now it was Lindemann's turn to be amazed, since Schrödinger was no more Jewish than he was. He couldn't imagine that a distinguished professor from Berlin who was still treated with respect by the Nazis could be thinking about emigrating.[9] Finally, Fritz London left Germany and lived on a stipend from Imperial Chemical Industries in Oxford, while Lindemann got the bonus of having Schrödinger temporarily in his laboratory. Like so many others, Heinz London and his boss, Simon, also went into exile and headed straight for England.

The Americans at this time were accepting as immigrants only those scientists who had already made a name for themselves. The British, on the other hand, while nominally following the same policies, were receiving many more scientists, especially the young, and they would have no reason to regret this situation later. The scientific refugees brought to England a wealth of ideas and the strong traditions of German science. For low temperatures it is no exaggeration to speak of a renaissance. Franz Simon's work on helium, for example, was remarkable, and he was rewarded with knighthood after the war—he became Sir Francis Simon. In France research groups developed only around a few famous names, in spite of the fact that the Popular Front, to its credit, had created a special funding organism for scientific research, the predecessor of today's National Center for Scientific Research (CNRS), similar in some respects to the National Science Foundation in the United States. In fact, France provided almost no help at all to scientific exiles. Given that many foreigners who did seek refuge in France were interned the moment war was announced and then handed over to the Germans after the debacle of 1940, the scientists were no doubt better off this way.

Heinz London lived for two years with his brother and sister-in-law in England. Fritz had close contacts with the cryogenicists at Oxford,[10] and together Heinz and Fritz thought about superconductors. Their articles no longer appeared in *Zeitschrift für Physik* but, rather, in *Proceedings of the Royal Society*. The scientific literature in low-temperature physics written in German would simply disappear.

It was here in England that the Londons worked on the famous London equation. In contrast to Becker, Heller, and Sauter, whom they cited, the Londons started off by making the Meissner effect a postulate of the theory. As discussed earlier, they required that $B = 0$ and $E = 0$ inside a superconductor. The Londons considered, for example, a current circulating around the surface of a sphere, perpendicular to the external field. The current of superconducting electrons, they said, would create an "induced" magnetic field opposite in direction to the external field but with the same magnitude. The net magnetic field inside would thus be zero, and the previous history of the superconductor didn't matter. Nothing they did contradicted Bloch's famous theorem, because a superconductor in a magnetic field was not at its minimum possible energy. Remember that Bloch had said that the most stable state of a superconductor, the minimum-energy state, was one in which there was no current. The Londons postulated that energy was supplied to the superconductor by the magnetic field, lifting it above its minimum energy. The bottom line is that, for superconductors, the Londons added a fifth equation to the traditional four of Maxwell (in which they eliminated all terms depending on the electric field, because the electric field is zero in the superconductor). The new equation stated that the curl of the current (a mathematical operation related to the derivative) was proportional to the magnetic field. The proportionality factor had the dimensions of length, and it was, in fact, just the penetration depth λ_L of the supercurrents.

Fritz London's stipend in England came with no long-term guarantees; he had to find a stable position.[11] In 1936 he briefly held the title of master of research and then director of research at the Institut Poincaré in Paris before he returned for a short period to England. To be eligible for a professorial position in France, he was required in 1937 to defend a thesis for the *docteur ès sciences physiques* degree. Another thesis! But the Paris lectures of 1935 that we've mentioned, "A New Conception of Superconductivity," provided the title and material for his successful defense. All for naught; at least no position was forthcoming. And so, in 1939, he finally accepted a professorship in theoretical chemistry (chemistry again) at Duke University, a position he occupied until his death in 1954.

Six years later his wife, Edith, wrote: "Honors came very late in Fritz London's life. . . . The political upheavals of the last few decades inevitably influenced the course of [his] life. The task of adjustments to new countries, new cultures, was a highly demanding one."[12] In her acceptance speech on the occasion of the posthumous award of the Lorentz medal to her husband, she said: "These simple, modest words give barely any indication of the heavy demands made on this dedicated life, demands of great courage and untiring efforts, of valiant persistence in defense of new conceptions, some of which were first received with skeptical attitudes ranging from the 'wait and see' of mild suspicion to violent hostility and fanatical rejection." Beneath the bitterness we simply find a widow defending the memory of her husband.

Of all the physicists interested in superconductivity at the end of the 1930s, only one had an approach to the problem that, little by little, turned out to be similar to that of London. This physicist never had the chance to meet London, but he dominated the entire landscape of Soviet physics and engaged in lively arguments with London about the properties of liquid helium. His name was Lev Davidovich Landau.

CHAPTER 7

Not Your Everyday Liquid

The liquefaction of helium provided access to the world of very low temperatures. Successful production of liquid helium had required a substantial technical infrastructure as well as understanding the subtleties of the liquid-vapor equilibrium of helium. Nevertheless, the ideas that dominated this thirteen-year-long competition, from the day helium was found on earth until its liquefaction in 1908, were squarely in the mainstream of physics at the end of the nineteenth century. This was not the case for the study of the new liquid, whose surprising properties could not be understood without quantum mechanics.

Helium is indeed an unusual liquid, but the scientists who gazed upon it every day didn't see this right away. It took them many years to notice that a phase with radically new properties could be obtained simply by cooling the liquid, a phase with infinitely small viscosity, reminiscent of the zero resistance of a superconductor. Once recognized, many more years passed before they really understood what was happening.

From the moment he liquefied helium, Kamerlingh Onnes had tried to solidify it by cooling it in the usual way, pumping on the helium gas above the liquid. The liquid thus evaporates, and, with the liquid-vapor system thermally insulated, it cools down. Kamerlingh Onnes continued to try this direct method of solidification until the end of his life. As better vacuum pumps appeared on the market, they were quickly thrown into the battle, and lower and lower pressures were achieved. But nothing worked. Even with a temperature as low as 1K, corresponding to a pressure of less than one millimeter of mercury, the helium refused to solidify.

Kamerlingh Onnes gave it one last try in 1922. Twelve diffusion pumps were put into service simultaneously pumping on a particularly well-insulated cryostat.[1] Kamerlingh Onnes got to 0.83K, but he didn't get solid helium. Once more in the history of the rare gases, the phase diagram was playing tricks — it turned out that helium could not be solidified simply by pumping; the liquid has to be cooled while the pressure is simultaneously increased. This required a completely different setup, and it was finally Keesom, Kamerlingh Onnes's successor, who succeeded, by applying

150 atmospheres of pressure at a temperature of 4.2K. The result seemed important enough that he sent off a quick telegram to *Nature* to announce the discovery.

Something interesting *was* happening in Kamerlingh Onnes's futile attempts to solidify helium, but no one was paying close attention. Each time he reduced the pressure above the helium, he observed that something apparently bizarre occurred when the temperature passed through 2.19K—the cooling slowed down considerably even though the pumping was constant. At first he blamed the thermometers, but it soon became clear that the thermometers were working properly. At 2.19K something was indeed anomalous in the liquid helium. To become really convinced of this would have required only a careful look at the helium, an easy thing to do. As we saw earlier, the helium had to be lit from below to see the interface that separated the liquid and gas phases and signaled the presence of the liquid. The glass cryostats in Leiden had a narrow observation slit that was perfectly suited for observing the boiling at different temperatures. Apparently, however, no one looked closely.

Since 1911 Kamerlingh Onnes had been studying how the density of liquid helium depends on the temperature. When he gave up pumping on the liquid helium, he returned to his studies of the physical properties of the liquid. He found a maximum in the density just at 2.19K; just like water at 4°C, helium occupies less volume at 2.19K than it does at both lower and higher temperatures. When ice melts, the molecules of the liquid are more or less in the same position relative to their close neighbors as the molecules of the ice. This local order disappears as the temperature increases and the volume decreases; at 4°C, however, the volume starts to increase. Nothing like this happens in helium. Helium molecules consist of but one atom, and they have only weak interactions with one another, so it is hard to imagine that they could squeeze together so much as the temperature approaches 2.19K.

The density maximum in water at 4°C signals the phase transition at 0°C. Kamerlingh Onnes might well have wondered whether the density maximum in helium carried the same message. A phase change, he thought, would be accompanied by latent heat and so, in 1926, benefiting from the visit of an American physicist, Leo Dana, Kamerlingh Onnes embarked upon the delicate measurements of the temperature dependence of the latent heat of vaporization of helium. Fix the temperature, measure how much heat has to be supplied to evaporate one gram of helium, then change the temperature and do it again. The results were disappointing—the amount of heat hardly varied with the change in temperature. The density maximum seemed not to be the harbinger of a phase change, at least not a traditional phase change. Still, such anomalous behavior had already been encountered in the transition between the normal state and the superconducting state of a metal.

Specific heat measurements are carried out with the same technique for both liquids and solids, and so data on specific heats had been obtained for helium at the same time as they were obtained for superconductors. Nevertheless, the results for helium were published only after Kamerlingh Onnes's death in 1926 and, even then, only for temperatures above 2.5K. Sure enough, the data for lower temperatures were also available, but the measured values were so large that the experimenters didn't believe their measurements and didn't dare publish them. In fact, of course, the dramatic increase in the specific heat was just another indication that something strange was happening in liquid helium near 2.2K.

Along with the large values for the specific heats, we mentioned that the experimenters had also noticed that it was more difficult to cool the helium as they approached the magic 2.19K, even though they were pumping on the gas equally vigorously. But this was really another indication that the specific heats were indeed very large there. If the specific heat is large, it means that more heat has to be taken away to lower the temperature by a fixed amount. With a fixed pumping speed, they were taking away a constant number of helium atoms per unit time as they changed the temperature, that is, a constant number of grams of helium. But Dana and Kamerlingh Onnes had shown that the latent heat did not vary with the temperature, so the constant mass of helium being pumped away meant that a constant amount of heat was leaving the system. Thus, the temperature dropped more and more slowly as the specific heat increased.

It took six years after the death of Kamerlingh Onnes before Keesom, now in charge of research on helium, took up the specific heat measurements again, this time in collaboration with a German visitor, Klaus Clusius.[2] Their results were the same as the earlier ones, but this time they were published. The specific heat climbs rapidly toward very high values as the region of 2.19K is approached, and it falls just as quickly once the temperature drops again. A graph of specific heat against temperature looks just like a fancy version of the Greek letter λ (lambda), and the temperature of 2.19K was baptized the *lambda point*.

Liquid helium undergoes a phase change at 2.19K. This much was now accepted, because different physical properties showed some kind of unusual behavior at this temperature. Thus, one now speaks of helium I above this temperature and helium II below, the names later given to the two forms of helium by Keesom. He noticed that the pressure required to solidify helium II remained constant as the temperature dropped below the lambda point. The educated physicist of the early days of the twentieth century would immediately think of an equation in thermodynamics known as the *Clausius-Clapeyron equation*: the variation in the liquid-solid equilibrium pressure is directly related to the difference in entropy between the two phases. For solid helium and liquid helium II the pressure stayed constant from the lowest temperatures achievable up to the vicinity of the lambda point. Keesom, good thermodynamicist that he was, immediately deduced that the difference in entropy between the two phases was zero. Helium II was just as ordered as a solid.

This result was at once paradoxical and reassuring. Reassuring because, after all, if helium II was just as ordered as a solid, that would explain its existence at very

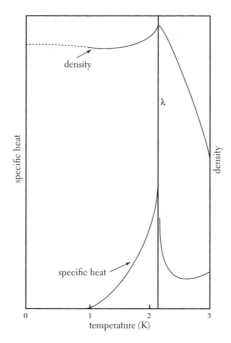

The λ point in liquid helium. When helium 4 is cooled at atmospheric pressure, it liquefies at 4.2K and then passes from the normal liquid state to a superfluid state at 2.19K. This transition is marked by a singularity in the specific heat, a narrow region in temperature in which the specific heat becomes exceedingly large. The solid curve shown, which represents the value of the specific heat as a function of temperature, has the shape of the Greek letter lambda (λ), so 2.19K is customarily called the lambda point. The mass density (dashed line then solid) shows a small anomaly but no discontinuity at this temperature.

low temperatures. Paradoxical because no one knew quite what to make of a liquid that was just as ordered as a solid. Analogous structures had been observed in certain organic substances at the end of the nineteenth century, especially by the French crystallographer Charles Friedel, the patriarch of a line of scientists who are still with us. These are liquid crystals, fairly commonplace these days, and used, for example, to display the time on a watch face. Keesom thus hypothesized that liquid helium II was a liquid crystal.

In a liquid crystal the molecules are free to move around relative to one another but nevertheless have a spatial organization similar to that of a crystal; this can be observed in optical experiments. In "nematic" liquid crystals all the molecules throughout the volume of the liquid crystal are oriented along a single axis. The molecules in a given plane of "smectic" liquid crystals are oriented parallel to one another, but the orientation varies randomly from one plane to the next. Finally, "cholesteric" liquid crystals, the ones used in watch faces, are smectic crystals in which the orientation of the molecules changes by a constant angle from one plane to the next.

Soon after Keesom's observations, Herbert Fröhlich, another theorist who was very active in superconductivity, took up this geometrical point of view with a similar model. He suggested that N atoms of helium, whatever their phase, could spread themselves over 2N sites of a body-centered cubic lattice. Such a lattice can be considered as the superposition of two simple cubic lattices (with atomic sites at the corners) arranged so that the vertices of the cubes in one lattice are the centers of the cubes of the other. Helium I, Fröhlich said, would correspond to a random positioning of the N atoms over the 2N sites, which would engender a certain disorder. He suggested that helium II atoms occupied the sites of only one of the two lattices, thus increasing the order (and decreasing the entropy).[3] The idea was clever, but it was invalidated by X-ray experiments carried out by K. W. Taconis, one of the young physicists in Leiden. A more or less regular structure for helium II should have yielded an X-ray pattern like that of an ordinary crystal, but nothing of the sort was observed. So, there was no crystalline order in helium II, even though the thermodynamic measurements demonstrated that helium II had to be ordered.[4]

It was not until 1933 that the Toronto group (Mc Lennan, Smith, and Wilhelm) finally published what, like the purloined letter of Edgar Allan Poe, every cryogenicist had under his nose but had never noticed. When liquid helium passed the lambda point, "the aspect of the liquid underwent a very sharp change, boiling stopped instantly. The liquid became very calm."[5]

Above the lambda point helium, like all boiling liquids, is turbulent, with large bubbles being created and exploding throughout its volume. Below the lambda point there are no big bubbles. For anyone, even a neophyte, who's had the chance to observe it even once, it's a striking and obvious phenomenon, and so it's incomprehensible that such a unique characteristic was not recognized for so long. Casimir sees this as a clear failure of the primacy of measurement, the credo of Leiden. Mendelssohn, being more conciliatory, suggests that the very enormity of the discovery prevented its happening earlier. Of course, it's quite possible that someone observed this transformation before the Toronto group but decided not to publish it because it was purely qualitative.[6] Even so, we would still have to understand why this phenomenon was never considered as a clue in trying to build a theory of helium II. This collective blindness to the evidence shows very clearly that "sensation" alone, in the sense of the philosopher-physicist Ernst Mach, is insufficient. Qualitative observations can be important only if they are made by an "informed" observer.

Before proceeding, we must compare the parallel histories of helium and superconductivity. The true nature of superconductivity, that is, the Meissner effect, was revealed many years after the discovery of superconductivity. The gap is comparable to the delay between the first liquefaction of helium and the discovery of the

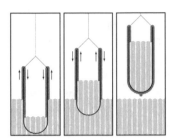 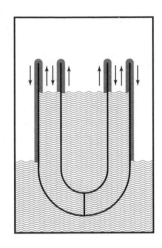

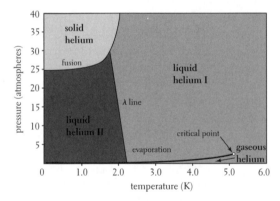

Top: The strange behavior of superfluid helium. Superfluid helium has zero viscosity and is perfectly "wet." The latter means that it will cover the entire internal and external surface of its container with a thin film; its zero viscosity means that the film will slowly flow, as shown by the arrows in the figure. The helium thus escapes from its container and will continue to flow as long as the level inside is higher than the helium level outside. Flow stops when the two levels are equal.

Bottom: Phase diagram of helium 4. In contrast with all other gases known previously, it seemed that helium 4 would not solidify. In fact, it was finally found that it would become a solid but only at pressures larger than 25 atmospheres. It thus has no *triple point* at which gas, liquid, and solid coexist. The density of helium hardly changes at the lambda point, and the lambda point depends only slightly on the applied pressure.

absolute transparency of liquid helium below the lambda point. For helium, however, physicists were quickly convinced that there was a phase transition. This was much more difficult to accept for superconductivity, at least for the experimenters. The difference in attitude is easily comprehensible. Classical methods for investigating phase transitions were applied to liquid helium, even if the temperature regime was extreme. In fact, most low-temperature research was devoted to looking for phase transitions and characterizing them by their thermal and structural properties. For superconductivity the situation was much more complicated because, as we have seen, everything depended on the observation of perfect diamagnetism demonstrated by the Meissner effect, an electromagnetic property and not a thermal one. In both cases, however, understanding the nature and organization of the new phase did not come easily.

The descent to absolute zero is accompanied by a decrease in thermal agitation and its associated disorder. The closer a material is to absolute zero, the more it is ordered. The models proposed to explain the properties of the superconducting state and those of helium II were first of all exclusively inspired by a geometrical approach to the organization of matter. Such an approach had been astonishingly successful in crystallography, including the physics of low temperatures at that time, in which most of the experiments investigated changes in the symmetry of crystals as the temperature changed. The number of setups designed to take X-ray photos at low temperatures was much larger than the number of cryostats in service during the 1920s and 1930s. Unfortunately, for helium and superconductors alike, the idea that a phase transition corresponds to a new geometric symmetry turned out to be without foundation.[7] This had been known for some time for superconductors; for helium II it was determined from the X-ray studies mentioned earlier. Yet the fact that no order appears in ordinary geometric space does not imply that no order of any sort appears in the transition. This would become apparent many years later.

Order does appear in these phase transitions, but this order is not visible as a different arrangement in ordinary three-dimensional space. This remark is not as complicated as it sounds. To represent the motion of free electrons in a metal, it's not helpful to consider their positions in ordinary space. Even if we could look at all of them at once, we would see nothing but frightful disorder, with electrons scattered randomly and moving in all directions. There is, however, another way to represent electrons in a metal, by the components of their velocities in three perpendicular directions. At absolute zero all the velocity vectors lie within a sphere (called the *Fermi sphere*) whose radius is the maximum allowed velocity. This representation in velocity space looks much simpler than the representation in ordinary space. Perhaps the reader can now imagine that helium II and superconductors can be ordered in this space even if they are not ordered in ordinary space.

Theory at the time had not yet reached such understanding. Yet Leiden's monopoly in cryogenics had been broken. It was in the new cryogenics centers that the understanding of helium advanced, especially in Cambridge and Moscow, in Cambridge in spite of Stalin and in Moscow because of him.

CHAPTER 8

The Russian Cold

Landau, Shubnikov, and Kapitza, the Russian trio that dominated low-temperature physics in the Soviet Union from its very beginning, were each caught in the turmoil of Stalinism like millions of others. In the unique circumstances created by the persecution of many prominent Soviet citizens, they continued their scientific work and obtained important results. The romantic with a taste for heroic stories will praise their courage and sacrifice. Rather than feed this tradition, however, it is perhaps more useful to remark that scientific progress is not a proof of social progress.

At the beginning of the 1930s two generations of physicists coexisted in the Soviet Union. The first consisted of the survivors of the old czarist intelligentsia who were not exiled. The second had been created by the Communist Party at its XVI Congress in July 1930, when it launched a "socialist offensive on the science front" and named "red experts" to take over the leadership of science and technology. These red experts were the sons of workers, selected primarily for political reasons. A direct witness, Alexander Weissberg, recounted in 1950:

> These young people had started out by taking courses in workers' colleges, and then they were dispatched into the universities; it was up to them to acquire the training that would make them real scientists. The designated ones were not always the best students, but the party leadership believed that this policy would have great results. They felt that among the vast proletariat unleashed in the schools the really gifted ones would not remain hidden for long. Up to now [Weissberg was speaking about 1937] that has not happened.

The scientific offensive was not only ideological, a denunciation of "Menchevist idealists" or the beginning of the Lysenko affair, but also budgetary. Indeed, at the start there was little money. A young woman, a student at the school of physics and mechanics of Petrograd, described one winter among many, the winter of 1921–1922:

> There was practically no fuel, and the classrooms were not heated. This forced the schools to close from January to April 1922, and the second semester was postponed

until the summer. Occasionally some classes for physics and mechanics students were held in the school offices, which were not very big. In the middle of the offices a brick stove had been set up, with a chimney that exhausted directly through a small window. One student had to get there early to light the stove. Since the wood was green, we often sat amidst thick smoke. In the autumn of 1922, conditions improved slightly.[1]

After these years of poverty, the new regime made a big effort to establish laboratories and new institutes. J. D. Bernal, the English crystallographer, visited the USSR at this time and expressed his strong enthusiasm for the Soviet program of scientific activities. It inspired him to write his famous book, *The Social Function of Science*, which was first published in 1939. Nevertheless, not a single high-class "proletarian" physicist was trained this way, although the USSR was not lacking in scientific talent, including A. Ioffe, a lively representative of the old generation, Y. Frenkel, I. Tamm, Landau, and Kapitza. Among the new institutes Kharkov would become very important in the history of superconductivity for a number of years, because of the simultaneous presence within its walls of Shubnikov and Landau.

In order to foster contacts between the most talented of the young Soviet physicists and scientists abroad, Narkompros, the People's Education Commission, organized a system of fellowships for stays in other countries. Given the difficult conditions of Soviet life during this period, it is not hard to imagine what these fellowships represented for their young beneficiaries. Nothing like them had existed before; the shortest visit to a foreign country had required vast treasures of imagination. This was true, for example, when Piotr Kapitza left the USSR in 1921. His father was a czarist general, a civil engineer, and his mother, the daughter of a general, was an intellectual, like Landau's mother. (Scientific biographies have an annoying tendency to talk only about the talents of the fathers.) Born in 1894, Kapitza, after secondary school in Kronstadt, obtained his electrical engineering degree at the Polytechnic Institute of Petrograd in 1918. During the turmoil of the revolution he taught in Petrograd and lost his wife and two children to famine and epidemics. At the same time, he wrote a half-dozen scientific articles. Abram Fedorovich Ioffe, a renowned scientist who had participated in the famous Solvay conferences in Brussels, was director of the institute and took Kapitza under his wing. Seeing his distress, he managed, apparently with the help of Maksim Gorky, to get Kapitza an exit visa from the Soviet Union. Kapitza got to England and, not without difficulty, convinced Rutherford to take him on as a thesis student in Cambridge in nuclear physics. (After Dewar's death there was practically no low-temperature physics in England).

For Lev Vassilevich Shubnikov, who was six years younger than Kapitza, his first years were less troubled, in spite of the civil war; he got involved in low-temperature physics right from the start. Like Kapitza, he studied at the Polytechnic Institute of Leningrad (the former Petrograd). For his 1926 diploma in condensed-matter physics, he tried to produce single crystals by solidification of molten metal. He wanted to determine their deformation under mechanical constraints. Kapitza himself had already helped to perfect the technique for producing the crystals. Benefiting from a Narkompros fellowship, Shubnikov and his physicist wife, Olga Trapeznikova, a childhood sweetheart, then went off to Leiden for their stay with de Haas.[2] Leiden's reputation was sufficient that the young fellows had no trouble justifying a stay of several years, so long as they made regular visits to the Soviet consulate in Berlin to take care of visa problems. Someone from Leningrad was not exactly in completely foreign territory in Leiden, since Ehrenfest, the successor to Lorentz, was a professor there. Ehrenfest was Viennese and a former student of Boltzmann, but he had married a native of St. Petersburg[3] and taught there from 1907 to 1912.

In Leiden, as we know, Shubnikov and de Haas improved the technique for producing single crystals of bismuth and obtained very high-quality samples, given the state of the art at the time. (Kapitza had occasion to visit them in 1928 on his way from Cambridge.)[4] Their work resulted in what we call today the *Shubnikov–de Haas effect*. But we're not concerned so much with this discovery as we are with the style of working that Shubnikov picked up there. The Dutch had an absolute mastery of low-temperature measurement techniques, and Shubnikov would establish his reputation as a physicist who was very attentive to the quality of his samples.

Shubnikov and Trapeznikova returned to the Soviet Union in 1930, while Kapitza stayed in Cambridge under Rutherford's patronage. Kapitza also received a stipend of two hundred pounds sterling per year from the Soviet government as well as his expenses for traveling to the Soviet Union to act as a consultant. This had all been arranged in April 1929, with L. N. Kamenev, the official at Narkompros in charge of scientific affairs. Upon his return to the Soviet Union, Shubnikov met Ioffe, the great organizer of Soviet physics in this period. Ioffe proposed that he create the first low-temperature physics institute in the Soviet Union in Kharkov, then the capital of Ukraine.

Teaching in the university in Kharkov at the time was a young man of twenty-two, born in 1908 in Baku. He had started his studies at this university at fourteen, received a diploma from the University of Leningrad at eighteen, and then done his thesis work with Ioffe in Leningrad, where he met Shubnikov and Trapeznikova. This young man had also been a Narkompros fellow for eighteen months, in the course of which he met Pauli, Dirac, and finally the man he considered his teacher, Niels Bohr. The young man was Lev Davidovich Landau, one of the most influential theorists of the twentieth century. Soviet physicists venerate his memory as no other.

When it comes to the Soviet Union, even after its collapse, it is difficult to distinguish well-established facts amid a fog of allusions, omissions, and lies. The problem is particularly acute for information about highly placed scientists. For Landau the problem has been resolved quite simply: denial. Myriad sources exist—written speeches paying tribute to the master, descriptions by his students and colleagues, and the like. Nevertheless, the only Soviet work devoted to Landau that is widely available assures us on page 4: "In the private life of this great scientist, there were no striking events that might be interesting to the public."[5] Nothing noteworthy, indeed, so long as a year in the gulag, for example, is considered an unimportant detail in a man's life.

Landau insisted that he not be considered a prodigy, but he was well aware of his talents nevertheless. At eighteen, even before completing his diploma studies, he published his first paper, in which he discussed issues about the new quantum mechanics. Already in his second paper, he made an important contribution.[6] His stay with Bohr in Copenhagen, where the atmosphere was strikingly free, obviously did not increase his respect for hierarchies, even though he was at the time a dedicated Marxist, like many of the young Soviet intelligentsia. When he returned home, Landau went back to work in the Physical and Technical Institute of Ioffe in Leningrad, as expected. Ioffe was always interested in concrete questions, and he proposed to Landau that he think about the feasibility of making an electrical insulator out of thin layers of molecules. Insulating materials were expensive, so a cheap alternative could be very attractive. Landau didn't worry about economics. He quickly showed in an article in *Zeitschrift für Physik* that the project was theoretically impossible,[7] and later experiments proved him right. Ioffe, however, was offended. Later, when Landau was making a presentation on another subject, Ioffe remarked that he didn't see any value in what Landau was doing. Landau didn't hold his tongue: "Theoretical physics," he replied, "is a complicated science that not everyone can understand."[8] With this outburst Landau left for Kharkov, where he met Shubnikov and Trapeznikova.

In Kharkov, Landau revealed his talent as a teacher and began to write his famous *Course in Theoretical Physics*.[9] At the same time, he also carefully answered the questions that the experimenters in Shubnikov's lab posed to him. Almost involuntarily, Ioffe had thus created in Kharkov one of those rare situations in which theorists and experimentalists attack the same issues. It was the same cocktail that had succeeded so well in Leiden, but, otherwise, conditions in Kharkov were quite difficult.

In many respects what happened in Kharkov during the 1930s is an extraordinary condensation of Soviet reality at that time, when vast technical projects backed by huge expenditures were given free rein in a context of abject poverty. Famine beat down the country once more in 1931, the farms were collectivized in 1932, and millions of people died of hunger. It was only in 1934 that rationing of food was stopped; the ruble of 1933 was worth no more than a fiftieth of the gold ruble of 1928.

Kamerlingh Onnes, in spite of his cleverness and clear-sighted view of his goals, needed twenty years in Leiden to reach premier rank. Meissner made his leap much more quickly in Berlin, but there he profited from the close connection between the university and German industry. The Kharkov laboratory was built and became truly productive, at least in superconductivity, in less than four years. Twenty institutes were founded at the same time, of which physics was the most important. In fact, the laboratory was nominally created in 1929 before the arrival of Shubnikov.

Ivan Obreimov, a scientist of the previous generation, the one who had introduced Shubnikov to crystallography, had been named director. Yet it was a young nuclear physicist named A. I. Leipunski who became the real leader. Most of the physicists came from Leningrad, and Shubnikov was named head of the low-temperature physics section in 1931. The lessons of Leiden were put into practice right away. Glassblowers and mechanical engineers were hired as well as a number of foreign specialists who, within a few years and in spite of the general poverty, acquired and mounted a complete set of cryogenic equipment, including liquefiers.

The Physics and Technology Institute of Ukraine was divided into eight sections, including one of the largest centers of experimental physics in Europe. Weissberg, a remarkable scientist who has since been mostly forgotten, described the beginnings of the low-temperature laboratory, where he arrived in 1931.[10] Weissberg and Ruheman were not Russian, but they, along with Trapeznikova, are our documentary sources for Kharkov. This role justifies a few words about their lives.

Weissberg was born in Kraków in 1901; he was a citizen of Austria because Kraków at the time was part of the Austro-Hungarian Empire. His family emigrated to Vienna, where he received an engineering degree in 1926; he then taught in a technical institute in Berlin. By 1925 he had become a communist. Weissberg thus had both academic and industrial experience and, as a militant communist, experience also as a leader of men. It was Leipunski, the leader of the physics laboratories in Kharkov, who recruited Weissberg and put him in charge of the low-temperature test group and thus of applied cryogenics. The Communist Party had announced that every effort had to be devoted to production, so the institutes had to engage all their resources in the service of industrialization. Low-temperature techniques were very useful in the chemical industry for separating and purifying gases. The Russians had to familiarize themselves with these processes, which were not widely known, so that they could proceed with the construction of major installations for liquefying nitrogen. The cryogenics laboratory had to make the physical measurements and study possible technical designs. Weissberg was charged with recruiting other foreign specialists, one of whom was Martin Ruhemann.

Ruhemann studied in England, where his parents had settled before World War I. His father held the chair of organic chemistry at Cambridge, but he was treated so badly during the war that the family decided to return to their native Germany after the war, and Ruhemann was naturalized at sixteen. He worked and published papers with the Berlin cryogenicists, notably Simon,[11] and was thus broadly versed in gas liquefaction and low-temperature experimental physics. He left Germany before Hitler took power and arrived in Kharkov soon after Weissberg, who had met him in Germany and converted him to Leninism. Ruhemann thus found it useful, indeed opportune, to give up his German citizenship and become English. The British consul in Russia accepted his request immediately and without asking any questions because, as he explained: "You speak English like the boys in our public schools. That would be quite impossible for someone not brought up in England." It was his English passport that allowed Ruhemann to leave the USSR freely

Cryogenics in Kharkov.

when his work contract expired in 1938. He left not because he had changed his mind about communism but, rather, because, as we know, even ideological purity was no guarantee that a foreign expert could slip through the massive Stalinist purges that began in 1937.

Shubnikov and his team had to fill many empty rooms in Kharkov, and that did not happen without difficulties. The helium liquefier came from Holland, of course, but it didn't work; original parts were replaced with parts purchased in Germany, but it still didn't work. Meissner himself came to Kharkov with his engineer, but no good came of this effort, and his visit ended in a squabble. Finally, they got the liquefier to work, but it made only about a liter and a half of liquid per hour, a meager output for such a sizable machine. Their hopes then lay in the mini-liquefiers that Simon had invented and Ruhemann helped obtain. Finally, the physicists had enough liquid to set about their experiments, but they could never get started until nightfall because they had to wait until each liquefier had filled its cryostat.[12]

Shubnikov. This photo was taken by his wife in the 1930s.

Shubnikov's bibliography reveals no publications until 1934. The laboratory was just beginning, and it was up to him to see that instruments were assembled so that physics could happen there. This meant that he couldn't participate in the decisive experiments that might have led to the discovery of the perfect diamagnetism of superconductors, the effect that Meissner discovered in Berlin well ahead of Leiden. This is not to suggest that there were no discussions about superconductivity between the two Levs of Kharkov, Landau the thin and Shubnikov the fat, as they were known in the laboratory. At the same time, Ruhemann was writing a lengthy book on low-temperature physics. The preface, which we might suppose was the last part written, is dated December 1935. Ruhemann added to the final proofs a list of publications that had appeared in the interim before the book appeared in print in 1937. This text, which often refers to Landau's point of view, is thus a faithful and direct account of the reactions of the specialists to the progress of superconductivity during this period. In fact, Ruhemann kept abreast of the discussions between Landau and Shubnikov, even though the liquefaction apparatus was far removed from the low-temperature laboratories. Weissberg also mentions frequent discussions between the three of them about the possible applications of liquid nitrogen. They saw one another frequently after work. Ruhemann, coming from Simon's lab, was the perfect interlocutor for Shubnikov, from the school of Leiden. The merits of the two styles of work could be compared and assessed.

Shubnikov and his assistant, Y. N. Riabinin, found a way to improve the sensitivity of magnetic field measurements at this time. The coil needed to measure the magnetic flux felt by a sample in a magnetic field was initially simply a wire wound around the sample. The only way to change the flux through the sample was to change the magnetic field, which involved, for example, turning a knob on a rheostat. Doing so limited the precision of the measurement and did not permit slow, smooth variations in the flux. Today the coil is no longer part of the sample. The sample can be put into the coil and removed, back and forth, while the coil and magnetic field remain fixed. This movement, like that of a sewing machine needle plunging rhythmically into a cloth, makes it possible to observe weak variations in the signal as well as its time evolution. Shubnikov and his aide invented this technique for their own studies of diamagnetism in metals and alloys.

By comparing monocrystalline and polycrystalline samples, Shubnikov convinced himself that it was only in single crystals of certain metals that the Meissner effect was truly and completely taking place. For other samples the effect is not complete; some flux remains enclosed in the volume of the superconductor. Landau tried to understand this result. He quickly showed that, if flux is not completely removed from the superconductors, there must necessarily be alternating strips of normal and superconducting material lined up parallel to the magnetic field. This is what is called the intermediate state. Landau obtained this result simply by evaluating the minimum energy necessary for the superconductor to oppose the penetration of the magnetic field. But this explanation was not nearly so clear to the experimenters, in Kharkov or elsewhere, because it was difficult for them to reconcile the seemingly contradictory properties of superconductors that they were discovering. We have already seen that the significance of the Meissner effect was hardly understood right away.

The net result of Shubnikov's many experiments can be summarized in the following way: the Meissner effect can be observed in pure metals. In certain alloys, when the magnetic field begins to penetrate into the interior, nonzero resistivity is not its constant companion. As Ruhemann put it: "We are thus confronted by the fact that alloys, in contradistinction to pure metals, appear to have two threshold values, one at which resistance appears, and another, much smaller, at which the field begins to penetrate into the conductor." Shubnikov was not the first to study alloys, but he was the first to characterize their behavior. Ruhemann's reaction was like that of all physicists involved in superconductivity: he didn't think these experiments were particularly fundamental. Alloys were simply a poor approximation to pure metals, he thought, so it wasn't surprising that the Meissner effect was incomplete or imperfect. They all thought that alloys were a dirty approximation of the clean ideal of a pure metal, and Landau's interpretation of the intermediate state carried little weight against this prejudice.

This example shows that science consists of more than measurements. Shubnikov's experiments were superb; what was lacking was a framework for interpreting them. Work on superconductivity during this period was taking place in an epistemological context very different from that of condensed-matter physics. Until then the properties of solids that had been known for a long time, such as resistivity and specific heat, had been used by physicists as a storehouse of data that allowed them to decide on the validity of the quantum description of condensed matter. The experiments on superconductors suddenly brought in a wealth of new facts, and the experimenters were not sure whether or not they were pertinent to a fundamental description of the phenomenon. In Shubnikov's results, for example, there was no indication that the behavior of alloys would furnish a very precise verification of the behavior of the superconductors that are now called type II superconductors. Far from being dirty, a poor man's version of a pure metal, alloys turned out to represent the ideal example of a type II superconductor; it would take a long time to conquer the prejudice. Shubnikov's experiments were twenty years ahead of theory.

The work of Shubnikov was well known in Europe before the war, if only via Ruhemann's book. This was especially true in England, where a number of low-temperature physicists had taken refuge. Shubnikov's reputation had already been established by his work in Leiden, and physicists spoke of the Shubnikov–de Haas effect. All these Soviet articles on alloys and superconductors were published, in German, in the first issues of the review *Soviet Physics*; until then Soviet contributions had appeared only in German journals. It was Weissberg who thought of starting this Soviet scientific journal. Given the situation in the USSR, he had to go to Moscow to plead his case before N. I. Bukharin, the former president of the Communist International. The article by Landau on the intermediate state appeared in 1937; it was the last article on superconductivity that came from Kharkov. The name of Shubnikov would, with the help of the war, soon be forgotten.

Since its creation the Kharkov institute had been an island of liberty, to use the expression of Weissberg, who was even then a critic of Soviet life. A photo taken in 1934 shows the scientists discussed here, young and unconstrained; even with all we know today about life in the Soviet Union at the time, they have an air of gentle confidence in themselves. Is this because they were physicists, and physicists all over Europe seemed generally to be in their thirties and taking delight in the intellectual excitement of the time? More prosaically, might not these young Russians have cherished the conviction that, given the sacrifices of others, they were the spoiled children of the authorities and would be spared the fate of their compatriots? After all, were they not the new ruling class? Several months later everything changed.

Niels Bohr took part in a conference on theoretical physics in Kharkov in 1934 because Landau was there. This probably marked the high point in the life of the laboratory. The substantial funds available for science compared to the penury of the general population strongly impressed the visitors, who had little idea of the reality of Soviet life. The concern about long-term planning was in total contrast to what passed for support of scientific activities in France or England. As late as 1937, Victor Weisskopf followed Bohr's suggestion and asked his friends in Kharkov whether scientists fleeing the persecution of Jews in Germany might be welcomed at the institute.

In fact, the situation there had become quite tense. S. Davidovich had been named as the new director to replace Leipunski at the end of 1934. This was the year that S. Kirov, one of Stalin's top aides, was assassinated in Leningrad, signaling a new wave of repression. It was also the year that Kapitza, as usual spending his vacation in his native Russia, was held against his will and not allowed to return to England. Davidovich was corrupt. He put his entire laboratory under the control of the GPU (a predecessor of the KGB); to establish his own position, he used it to turn researchers against one another. He had only to declare certain projects as military to give them a high priority; he kept decisions secret to achieve his own ends. Thus was Riabinin set up against his mentor, Shubnikov. Another program that Davidovich had decided on was pursued against the advice of Landau. Davidovich then had Landau's student and close friend, M. A. Koretz, arrested. This time Landau and Weissberg were able to get around the GPU of Kharkov; Weissberg used his political talents to draft just the right letter directly to the central committee. Davidovich was fired and Koretz liberated — totally extraordinary events — and Leipunski was restored to his former position. The GPU would not forget this affront.

Landau continued to show his independence, which, according to F. Janouch, led Landau to oppose the rector of the university on a simple pedagogical issue — possibly a deliberate provocation. In any case he resigned his professorial position in Kharkov to go to Moscow. When the 1937 wave of repression, the great purge, was unleashed, Weissberg, Koretz, and Landau were arrested. We know what happened to Weissberg, because his estranged wife had had him meet the writer Arthur Koestler before they went to the USSR. As usual, Weissberg had convinced Koestler of the justice of the Communist cause. Weissberg's wife, who went to live in Moscow, was arrested before him. When Weissberg protested, she was freed, while he took her place in the hands of the GPU. When she was expelled from the USSR, she alerted Koestler to her husband's fate.

Koestler had been interested in science since his first years as a journalist. He was the first to tell Louis de Broglie that he had won the Nobel Prize, a scoop that won him an exclusive interview. Much later, when he was out of work, Koestler had even been sheltered in Kharkov by Weissberg. Koestler thus expended enormous energy to have Weissberg freed. Einstein and the Joliot-Curies petitioned Stalin on his behalf, to no avail. Weissberg's odyssey is well described in his book *The Accused*, published in 1951. After a succession of interrogations and considerable abuse, one night Weissberg was confronted by Shubnikov, who himself had been arrested on August 6, 1937, on his return from a vacation in the Crimea with Lev Landau. Weissberg had already twice confessed to having wanted to conspire to kill Stalin, and twice he had retracted his confession. Here is his account:

Torniev, the prosecutor, had in hand the confession that Shubnikov had just signed. He asked the prisoner to repeat his deposition. Shubnikov declared: "Weissberg entered our institute in 1931. He came from Germany. The Gestapo had signed him up to organize sabotage and espionage in the institute, and he tried to get me involved as well.

But I refused, because I had already been an agent of the central German spy network since 1924. Since then, we've worked in parallel, but we've had no contact with one another."

Shubnikov never dared to look at me while he murmured these words. What had they done to this strong and stubborn man? He seemed exhausted, even though his body showed no sign that he had been treated badly. Why had he put his signature on these crazy depositions? Shubnikov was without doubt a man of high integrity. At the Institute, he was esteemed for his energy and determination. He was inflexible. How had they managed to reduce him to this state? Torniev turned toward me:

"Do you confirm the depositions of Shubnikov, the accused?" I paid no attention to him, and instead demanded of Shubnikov:

"Lev Vassilievitch, are you out of your mind? How could you sign such things?
—You have no right to speak to the accused. You can only respond to my question: yes or no?
— OK. No. I don't confirm his deposition."

Torniev took up every question in detail. I said "No" to every one. He didn't try to pressure me. He wrote down my responses and gave me his report to sign. Then he buzzed for a guard. As he was taking Shubnikov away, I gathered up all my courage and cried out to him: "Shubnikov, for heaven's sake, retract it all for the judge!"

The guard, who was behind me, struck me sharply on my head. The interrogation was over.[13]

This is the last we know of Shubnikov. He was condemned by a troika on November 28, 1937, to ten years of solitary confinement. On June 11, 1957, his name was cleared, but it was claimed that he had died in the gulag on November 8, 1945. In fact, he had been executed shortly after he was sentenced. It was forbidden to cite his name, and it disappeared from all publications.[14]

When Shubnikov and Weissberg met in 1937, Landau, Koretz, Obreimov, and Leipunski had already been arrested, like more than 5 percent of the Soviet population. These were scientists of truly rare talent, but in the Soviet Union under Stalin it was not rare for scientists, whatever their talent, to be shipped off to prison camps.

CHAPTER 9

In Cambridge in Spite of Stalin, In Moscow Because of Stalin

The histories of both the Cambridge and Moscow laboratories were marked by Piotr Kapitza's personality and his (mostly unintentional) quarrels with the Soviet authorities. Having arrived for a short stay in Cambridge in the winter of 1921, Kapitza finally convinced Rutherford to keep him on, even though there was no opening. There's an old story—we can't vouch for its accuracy—that has Kapitza asking two questions of Rutherford. The first: "So how many people do you have here in the laboratory?" to which Rutherford replied, "Thirty." The second: "How accurate are your measurements typically?" "Three percent." And then, according to the story, Kapitza drew the conclusion: "If you can't measure to better than 3 percent, having one researcher more or less won't make much of a difference; he'll be lost in the noise."

Physicists love this sort of anecdote, in which their professional quirks get tied up with daily life. More seriously, we can try to imagine what it was about Kapitza that seduced Rutherford. In spite of their many differences, what they had in common was that both were emigrés (Rutherford came from New Zealand) who had to assert themselves in the midst of a society that, even in its intellectual circles, seemed paralyzed by tradition.

At first, since it was the business of the laboratory, Kapitza got involved in nuclear physics experiments. Soon he needed high magnetic fields to bend the trajectories of charged particles, such as alpha particles and electrons. Very quickly Kapitza became a pioneer in the new field of intense magnetic fields, and, by the same token, he found himself involved in low-temperature physics.

> The best way to create a magnetic field is to generate an electric current in a coil—the higher the current, the stronger the magnetic field. But high currents also cause the coil to heat up; the heat grows as the square of the current and is put to good use in a toaster or an iron. As the temperature of the coil mounts, its resistance also rises; this process eventually limits the strength of the magnetic field that can be attained. If the coil is cooled by

liquid nitrogen, however, the resistance is lowered, and higher magnetic fields can be reached.

Soon the intense magnetic fields were being used also for the study of the properties of solids at low temperature. As Rutherford's protegé, Kapitza was put in charge of the construction of a new laboratory at Cambridge, the Mond Laboratory. It opened in 1933, with Kapitza as director. At the same time, Kapitza became interested in the liquefaction of helium. Fascinated by engineering, he attacked this problem in his usual fashion, beginning from the beginning, or, as a physicist would say, "starting from first principles." He made a prototype liquefier and then perfected one that operated continuously, in which the helium gas, initially compressed, was released to do work; this provided the precooling. The liquefaction itself followed as the result of a Joule-Thomson expansion. The machine included some remarkable innovations, especially the use of the helium itself as the lubricant for the piston. This liquefier's great advantage for the user was that it needed only liquid nitrogen for the initial cooling; for the first time a large-capacity liquefier could operate without liquid hydrogen, which had always been a serious safety hazard. The principles of Kapitza's liquefier continue to be used in all modern high-capacity helium liquefiers.

With this development the United Kingdom renewed its ties with low-temperature physics, which had been abandoned upon Dewar's death in 1923. The arrival of other young refugee scientists extended this renaissance beyond Cambridge. Simon, Mendelssohn, and Kurti came to Oxford in 1934, while Fritz London found a position in Bristol.[1] The world of cryogenicists was small and well traveled, so all the newcomers were known to their colleagues before they arrived in England. None of these refugees was particularly famous, because they were all at the beginning of their careers; their contribution at this time owed much to their number. It was not so large, but just before the war, according to Kurti, Oxford had only eight low-temperature physicists, four of whom had their Ph.D. degrees. The emigrés thus constituted an essential part of the structure, and their work proved to have a decisive influence on their new laboratories — Oxford became a close replica of Breslau in both personnel and equipment.[2] The contrast between the late 1930s and the 1920s was particularly marked in the British laboratories.

Curiously enough, France experienced a long eclipse of cryogenics at the same time as its neighbor across the Channel. The two countries undertook no low-temperature research, even though the liquid air industry was strong in both places. Industry limited its research to methods of obtaining oxygen, nitrogen, and liquid air; companies refused to expend resources on the natural extension of such research, the physics of low temperatures. So, here we have powerful and prosperous companies doing no fundamental research. For them the 20K of liquid hydrogen would be for decades the lowest temperature they cared about. On the other hand, these industries expended considerable effort in legal battles over the patents for

Physicists at Kharkov. This photo was taken on the steps of the Physical-Technical Institute of Kharkov during Piotr Kapitza's visit in 1934. *From left to right, first row:* L. V. Shubnikov, A. I. Leipunski, L. D. Landau, and P. L. Kapitza; *second row:* V. Finkelstein, O. N. Trapeznikova, C. D. Sinelnikov, and Y. N. Riabinin.

liquefiers. Such behavior is reasonable if the thinking is strictly commercial, when maintaining market share and national monopolies are what count. The bottom line was and is still the only criterion for their development. When, after the war, the markets opened for low-temperature scientific instrumentation, the French company L'Air Liquide and its counterpart, the British Oxygen Company, hardly played a role. This attitude is similar to that of the coal companies, which paid no attention to carbon chemistry except when they were forced to.

Unfortunately, Kapitza's participation in the low-temperature renewal in England came to an early end. Since his arrival it had been his custom to return each summer to Russia, where his mother still lived. He used this opportunity to honor the consulting contract that he had signed with Kamenev in April 1929. To be sure that he could return to Cambridge, he took the precaution of getting his exit visa from Russia each year before he left England. In 1934, as usual, he sent off a letter to this effect but left before he had a response. In addition, Rutherford wrote to the Soviet ambassador in London for assurance that Kapitza would indeed be back in Cambridge in September.[3] There's little doubt that Rutherford considered Kapitza as his heir-apparent, in spite of the prodigious talent assembled at the Cavendish Laboratory.

Once in Russia, Kapitza did not stay put. In particular, according to Ruhemann, he went to Kharkov to see Shubnikov and stayed at his house.[4] Kapitza had come to

Russia with his second wife, Anna Krilova. He had met her in Paris, where she lived alone with her mother, who was separated from her husband, the mathematician A. N. Krilov. When the moment came to return to England, Kapitza tried unsuccessfully to get his passport; only Anna got back to England. There she mobilized Rutherford and Kaptiza's friends to help get him out of the Soviet Union. Dirac organized many protests, but the protestors ran up against a brick wall. The Soviet embassy even tried humor in responding to Rutherford that Russia would be happy to let Kapitza leave, if only Rutherford would take his place. Rutherford died in 1937 without having seen Kapitza again. From then on, and not only in totalitarian countries, scientists were often looked on less as members of a profession than as government agents.[5] It was only in 1966 that Kapitza was in Cambridge once more.

The relations between Kapitza and the Soviet authorities were always ambiguous under Stalin and, later, under Khruschev. Kapitza was confined to the USSR. Given the general economic conditions, he lived well but was treated badly. For several months in Russia, Kapitza was threatened and followed by agents of the secret police agency, the NKVD, which he complained about in a letter to Molotov on May 7, 1935.[6] To protest what was effectively house arrest, he didn't hesitate to go on strike, giving up all scientific activity for two years, nor did he avoid telling the leaders what he thought of their procedures. On May 14, 1935, he wrote once more to Molotov: "You've told me that you have lots of young Kapitzas and I'm sure you not only have Kapitzas but super-Kapitzas. But if you go fishing among your hundred and sixty million people, you won't catch a one with your methods. That's why you had to get one from England with Rutherford's help."[7] During one of these two winters Ruhemann paid him a visit and remembers that, right there in Moscow, "Kapitza was in his dining room drinking vodka like an English gentleman drinking his whisky. It was very bizarre."[8]

He was at a dead end. What was happening elsewhere in the Soviet Union revealed this only too well, even if visitors didn't always realize what was going on — a new series of purges had started at the end of 1934. Finally, Stalin offered Kapitza a golden cage, the equivalent of what Rutherford had given him. They would construct an institute for him in Moscow that he could design himself; he would be in charge of everything, including the administration. Thus did Kapitza set about a quest for land in Moscow, with car and driver, but he didn't like what he was shown.

Since 1933 the U.S. ambassador in Moscow was a colorful personality named by President Roosevelt, William Bullit. Coming from high society in Philadelphia, he got involved in Russian-American diplomacy at a young age. He had met Lenin in February 1919, on behalf of the U.S. government, although he wasn't authorized to negotiate with him. Bullit was favorably impressed with the ambitions of the revolutionaries, whose hold on power at the time was extremely shaky. Lenin told Bullit that, if the West would stop supporting the white Russians, Russia would give up its claims on Finland, the Baltic States, and part of Ukraine, and would recognize the czarist debts. These extraordinary concessions were rejected, and Bullit resigned. When he later returned to Russia he was thus a priori favorably inclined toward the

Soviet Union. By that time he had been married for several years to the widow of John Reed, the author of *Ten Days That Shook the World*, whose body was enshrined in the Kremlin wall. Yet Bullit did not understand that the situation was no longer the same. Stalin had replaced Lenin, and the enormous power he had concentrated in himself rendered any concessions useless. When Bullit brought up the question of the repayment of Russian loans once more, the atmosphere, which was bad enough to begin with, quickly soured. To make his attitude perfectly clear, Stalin refused to grant the Americans the land in the heart of Moscow that they had targeted as the site for their embassy.[9] You guessed it — Kapitza found this wooded site to his liking, perfectly suited for the construction of the Institute for Physical Problems.

As in Kharkov, a whole array of buildings was constructed to house scientists and technical staff and to provide space to carry out their research and development. The director's apartment, consisting of twelve rooms and a large terrace, was right next to the main building, where the laboratories were located. Still, there was the problem of equipment. Kapitza ordered or had constructed the best there was, both for the machine shop and for the cryogenics labs. The key piece of equipment, however, at least at the start, was the Cambridge liquefier, which proved more problematic. After an exploratory trip to Moscow by P. A. M. Dirac and E. D. Adrian, Rutherford was convinced that Kapitza would never return.[10] Nevertheless, he did not want to deprive his protégé and spiritual heir of the fruit of his own labor. Negotiations ensued between the two governments, yielding an agreement that was a little more complicated than usual. England agreed to sell to the Soviets, for a price, Kapitza's equipment from the Mond Laboratory. In fact, Kapitza received all the original equipment used in his high magnetic field experiments. What he did not get was his original liquefier and its auxiliary apparatus, which stayed in Cambridge. Instead, the English made a copy of the original and sent it off to the Soviet Union. In addition, they sent along the English technician H. E. Pearson, who used to work with Kapitza at the Mond Laboratory, to help assemble all the pieces in Moscow. Finally, David Shoenberg, who had been Kapitza's student in Cambridge, went along as well, to continue his work. Because of Stalin, Kapitza was thus obliged to work in Moscow. In spite of Stalin, the Mond Laboratory continued to be a leading center of cryogenics research.

According to Shoenberg, Kapitza was not very happy to learn that he had to leave behind his arsenal for a strong competitor.[11] Indeed, as we shall see shortly, his fears were well founded: Cambridge proved to be the direct rival of Moscow in exploring the properties of liquid helium. Kapitza's liquefier in Cambridge was so well designed that after six years of downtime during the war it was put back into service and used until 1949.

Kapitza's laboratory was a high-priority project and was quickly constructed and equipped. Ruhemann was present in Moscow for its inauguration in 1936, and he could serve as an interpreter for Pearson in his meetings with the official visitors who had come to see the cryogenics laboratories. Shoenberg arrived in the laboratory around this time. A photo that he took shows a simple and elegant building, with a portrait of Lenin mounted on the top left, and a large XX just above the entrance to mark the twentieth anniversary of the revolution. It was Shoenberg who later, in the 1950s, circulated copies of a translation of the article by V. L. Ginzburg and Landau laying out the phenomenological theory of superconductivity that we'll encounter later.[12] Schoenberg also was witness to the first drama that shook the Institute for Physical Problems.

It started with some good news. In February 1937 Kapitza received this brief note scribbled in longhand:

> To the director of the Institute of Physical Problems:
>
> REQUEST:
>
> I would like to be accepted as a scientist in your institute.
>
> 2/8/37 L. LANDAU

Kapitza agreed immediately. Landau was set up in one of the little houses lined up side by side in front of the institute. In the laboratory he encountered talented experimentalists and an atmosphere conducive for scientific work. Here, finally, he and his group of theorists no longer had to waste time in conflicts with the administration, even if Kapitza from time to time would let off steam in a tirade against theorists. Landau's merits were fully recognized, and he got all the space he needed. Before getting to this point, however, Landau was once more caught up in the Kharkov purges. No sooner was he set up in the laboratory than he was arrested. This came at the peak of his scientific productivity. In 1937, in addition to working on his treatise on theoretical physics, he published ten articles; he would never do more in a single year. Among them were several on the theory of second-order phase transitions in the *Journal of Soviet Physics* that proved to be the point of departure for Landau's activities in superconductivity and the study of helium II.

The circumstances surrounding the arrest of Landau came to light only in 1993, when extracts of his KGB dossier were published.[13] He was detained with Koretz, who had already been arrested in Kharkov. The two of them were in contact with Fyodor Raskolnikoff, a member of the opposition within the Party who had been relieved of his duties in the Soviet Union and sent to Paris in 1935. Together they were preparing a tract for the May 1 celebrations of 1938 that contained a violent denunciation of Stalinism from the Left. This tract turns upside down the image of the Landau of this period that we might have had, a man who, while he seemed to have been through a lot, had not appeared to be in open revolt:[14]

> Comrades!
> Our great October revolution has been sold off by cowards. Our country is overrun with torrents of blood and filth. Millions of innocents have been thrown into

prison, and no one can know when his own turn will come. Is it conceivable, comrades, that you cannot see that the Stalinist clique has veered off toward fascism? Socialism no longer exists except on the pages of newspapers dedicated forever to the lie. In his hatred for real socialism, Stalin is just like Hitler and Mussolini.

By bringing this country to its knees to maintain his power, Stalin has made it an easy prey for the beast of German fascism.

In spite of its virulence, this tract was itself not the principal accusation against Landau, according to the KGB dossier, but it gives us a better appreciation of the reaction of Kapitza, who showed a courage rare in such circumstances. That same day he wrote directly to Stalin:

Moscow, April 28, 1938

To J. V. Stalin

This morning L. D. Landau, a scientist of this institute, was arrested. He's only twenty-nine, but along with Fock[15] he is the most eminent theorist in the Soviet Union. His papers on magnetism and quantum theory have been cited often in the scientific literature both here and abroad. Just last year Landau published a remarkable paper in which he identified for the first time a new source of stellar energy. It gives a plausible explanation of the fact that the energy of the sun and other stars is not yet exhausted. Bohr and other eminent scientists predict a great future for Landau's ideas. There is no doubt that the loss of Landau for our institute, for Soviet science, and for science the world over will not pass unnoticed and that it will provoke strong protests. Of course, knowledge and talent, no matter how extraordinary, confer no right to break the law; if Landau is guilty, he has to pay the price. But, considering his exceptional gifts, I ask you to order that his case be very carefully examined. It seems to me that Landau's abrasive personality should be taken into account. He is quarrelsome and he likes to find others' mistakes. If he finds an error, he takes devilish pleasure in harassing the perpetrator, especially if it's an old and pompous gray head like our academicians. I have to say that it hasn't been very easy to get along with him at the institute, although recently he seems to have taken some of our remarks to heart and behaved a bit better. I forgive his jokes because of his extraordinary talents. But in spite of all these character faults, I find it difficult to believe that Landau could ever do something dishonest. Landau is young and he can still produce an enormous amount of science. No one is better positioned to judge this than another scientist; this is why I write to you.[16]

And so, in these extraordinary circumstances, Kapitza outlined a most exact description of Landau's personality. Landau had no illusions about the Soviet state; as he said to Shoenberg, "We invented concentration camps."[17] The letter had no effect.

This action by Kapitza, on this occasion as on many others later, makes hardly credible the rumor that he was a spy in Cambridge. The idea was that he was a member of the elite Apostles Club, there at the same time as the known spies Guy Burgess and Kim Philby. This rumor was repeated, without the least effort at documentation, by P. Wright, the retired chief of the British section M1 and the author of a book that the government of Margaret Thatcher wanted to keep out of print for other reasons.[18]

Kapitza's attitude seems even more remarkable if we remember that he was not only getting his institute established but was also engaged in intense experimental activity that led to important results at almost the same moment as Landau's arrest. His point of departure was a measurement of the viscosity of helium. One of the great disasters in a low-temperature experimentalist's day is to find a bad vacuum in his cryostat. Of course, it always happens when the experimenter had planned rather more fruitful activities for himself than repairing an essential piece of his apparatus. If there's a bad vacuum, quite obviously, there's a leak somewhere. Now, many of the essential properties of helium are due to the small size of the helium atom. The slightest hint of a break in a container, invisible to a microscope, is for helium a veritable boulevard where the helium can stroll as it pleases without encumbrance. There are even cases where the leak is apparent only below the lambda point, and the vacuum is tight above. One bit of advice: pay no attention to your colleagues who tell you how they managed to localize a leak even in such an extreme case. Change the cryostat! You'll understand why when you get to the end of this chapter. These "leaks that open up only when the temperature drops," explains Joseph Behar, the engineer at the superconductivity laboratory in Orsay near Paris, were well known historically: the Keesoms, father and daughter, were measuring the specific heat of helium II when they fell victim to such a leak. They were not second-rate physicists and quickly deduced that the viscosity of helium II must be smaller than that of helium I, since liquids with a low viscosity flow more readily.

Clearly, the viscosity was an interesting property and had to be measured. But Leiden no longer had a monopoly on cold. When measurements began in Leiden, they were started also in McLennan's laboratory in Toronto. This was the same McLennan who had arranged the famous trip of the superconductor immersed in liquid helium from Leiden to London while the supercurrent persisted. In addition, Cambridge and Moscow were now also in the race.

Both in Holland and in Canada a classic setup was used for viscosity measurements. A pile of disks separated from one another along a vertical axis is immersed in the liquid and made to oscillate around their common axis. Because the oscillation is horizontal, gravity plays no role. The oscillation is slowed by the friction between each disk and the liquid—the more the friction, the higher the viscosity, and the more quickly the oscillations are slowed down. The measurements in Leiden and in Toronto showed that in fact the viscosity of helium is somewhat lower below the lambda point than above. In addition, helium II was found to be a perfect heat conductor. This result is surprising for a liquid that doesn't conduct electricity. And, just to make the puzzle a little more difficult, the viscosity measurements seemed to contradict the observations of helium micro-leaks, which indicated that the viscosity of helium II was *much* lower than that of helium I.

J. F. Allen and A. D. Misener in Cambridge and Kapitza in Moscow independently of each other then tried a completely different technique to measure the viscosity of helium. When a liquid passes through a small orifice, such as a capillary or a very narrow slit, its flow velocity is much slower if its viscosity is high. Measured this way, the viscosity of helium below the lambda point turned out to be a million times lower than above. The two series of measurements were published in the same issue of *Nature* at the beginning of 1938. It seemed that the value of the viscosity depended not only on the type of experiment but also on the place where it was done. At this point Kapitza didn't get upset about the result and simply coined a new word to describe the phenomenon: *superfluidity*. The experiments showed that superfluid helium had neither viscosity nor turbulence.[19] None of the usual concepts of hydrodynamics applied. Decidedly, this was not your ordinary liquid.

Helium II has a number of other astonishing properties. It climbs the wall of a container in which it is enclosed and might even empty the container. It is also an infinitely good conductor of heat. All these properties were discovered between 1936 and 1938. The viscosity of helium II is, in fact, rigorously zero, just like the electrical resistance of a superconductor. This analogy, both evident and stimulating, posed an additional problem. Up until then, the prevailing opinion was that the enigma of superconductivity could be resolved only by finding some mechanism involving the electrons of the metal to explain the disappearance of resistance and the Meissner effect. The superfluidity of helium, however, could surely not be explained by any property whatsoever of electrons. Superfluid or not, helium remains an insulator and has no free electrons. Was there, beyond the superficial analogy, something in common between the two situations? Not surprisingly, the theorists who had been interested in superconductivity immediately began their attack on superfluidity. The experiments had been slow to demonstrate the phenomenon, but the theoretical successes came remarkably quickly, at the cost of bitter polemics.

One year had gone by since Landau had been arrested. Kapitza could not resign himself to the loss of such talent. This time he wrote to Molotov:

To V. M. Molotov

April 6, 1939

In my recent studies of liquid helium near absolute zero, I succeeded in discovering several new phenomena, which promise to shed light on one of the puzzles of contemporary physics. I expect to publish some of these results in the next few months, but I need some theoretical assistance. In the Soviet Union, it is Landau who has the most expertise in this domain; unfortunately, he's been in prison for an entire year.

> *All this time, I've hoped that he would be released because, to speak*
> *frankly, I can't believe that he has committed a crime against the state. My*
> *conviction is based on the fact that Landau, as a young theorist, brilliant*
> *and ambitious, who at the age of only thirty is already well known all over*
> *Europe, must be totally absorbed in his scientific work. He could hardly*
> *have the motivation and find the energy to undertake a completely different*
> *activity.*

Having shown that the accusations made no sense, Kapitza followed with several requests in the spirit of the times:

1. *Is it not possible to convince the NKVD that they should accelerate their investigation of Landau's case?*
2. *Wouldn't it be possible to use Landau's brains for scientific research while he's in the Butryki prison? I've heard talk of such a procedure's being followed for some engineers.*[20]

This intervention proved decisive: Landau, almost dying, was freed. Kapitza had to provide a guarantee to Beria, the head of the NKVD:

> To L. P. Beria
> Peoples' Commissar for Internal Affairs
> MOSCOW, APRIL 26, 1939
>
> *Via the present note, I request that Professor Lev Davidovich Landau be released from prison with my personal guarantee. I guarantee to the NKVD that Landau will not engage in any sort of counter-revolutionary activity against the Soviet government in my research institute and I will also take all necessary measures in my power to ensure that he does not engage in any such activities outside the institute. If I hear about any remark that Landau makes that might be prejudicial to the Soviet State, I will immediately inform the appropriate organs of the NKVD.*
>
> P. KAPITZA[21]

In 1940 Landau, fresh from prison, published an article on phase transitions.[22] It is in this article that he discusses the basic ideas that would permit him, in collaboration with Ginzburg, to propose what physicists call a *phenomenological theory of superconductivity*. Ehrenfest had already tackled this subject. Now, however, this approach to superconductivity had additional justification from the discovery of superfluidity, which from the beginning looked to some like a phenomenon analogous to superconductivity. But superfluidity was not yet understood.

CHAPTER 10

Superfluidity
Theories and Polemics

Even before World War II, physicists found it difficult to change their research focus from superconductivity to superfluidity. Helium required so much specialized equipment that experimenters very quickly decided to study one or the other, never both at the same time. The scientific heritage of Kamerlingh Onnes had led to a separation between helium and superconductivity research in Leiden; now this division became entrenched all over the world. In addition, the political atmosphere made everyone nervous. Mendelssohn, who had arrived in Oxford in 1933, recalled: "Progress in research was punctuated by alarming news from Germany. The shadow of war was beginning to spread over Europe and scientific research was becoming a race against the clock." [1]

Exchanges between specialists in helium and those in superconductivity were thus limited to a small circle of theorists, even though the concepts developed for helium turned out to be useful also for superconductivity. In today's physics texts it's taken as obvious that superconductivity in metals and superfluidity in helium are closely related phenomena. This was not what the majority of physicists believed. They were creative, but they thought of themselves as doing "normal" physics, exploring a field whose outlines and patterns were well established. As it turned out, dramatically different concepts would be required, a rupture of the paradigm, "revolutionary" science in the sense of Thomas Kuhn. [2] The paradox represented by the apparent inconsistency between different measurements of the viscosity of helium II was the real trigger for understanding the nature of this fluid. The discrepancies between the results did not refute any theories, but they posed particularly clearly a problem that had to be solved.

Fritz London was the first to make serious progress on this issue, just as he was for superconductivity. When the entire problem had been resolved, he was able to describe in just a few words the obstacles that had faced him and his colleagues: "Liquid helium seems to be too simple to have a λ-point! Indeed, it must be a peculiar kind of ordering process by which liquid helium cooled below 2.19°K rapidly loses all its entropy of liquid disorder without going into the solid state. The expression 'liquid degeneracy' has been suggested [Simon, *Zeit. für Physik* 41 (1927): 808] to give a name to this mysterious process. The mere naming of it, however, only served to emphasize the mystery." [3]

Like other physicists involved with helium, London had first played with the idea that the increased order apparent in helium below the lambda point was similar to the order among atoms in a crystal. Once he had conceived the idea of a macroscopic quantum order for superconductors, however, he came back to take a more serious look at helium. He soon abandoned this geometrical approach and instead took up its opposite, the perfect gas. His idea was to resurrect the old work of Einstein on gases that obeyed Bose statistics (and not Fermi-Dirac statistics, as electrons do). [4] In a hypothetical Bose gas the number of particles that can occupy a single state is not limited to two (as it was in our Fermi-Dirac model of couples filling rooms in the apartment building) — in fact, there is no limit at all. Einstein had shown that cooling a Bose gas should produce a most interesting effect: the gas would condense, not in the form of a liquid or a crystal but, rather, as a surprising new form of matter in which all the confined particles would move as slowly as quantum mechanics allows.

London suggested that the change of state of helium at the lambda point was just such a Bose-Einstein condensation; no other example had ever before been observed. This suggestion was published in 1938 shortly after the contributions of Kapitza and of Allen and Misener, all in volume 141 of *Nature*. Assuming that helium was an ideal Bose gas, London calculated that condensation would start at a temperature of 3.13K. Below this temperature some particles would begin to be found in the lowest state, that is, with their minimum speed.

We have already seen with the theory of Drude for the electrical resistance of metals that a poor model can sometimes yield nature's results. Here we have the opposite. The numerical agreement in London's calculations was by no means perfect, but his basic ideas would, nevertheless, turn out to be correct. The number of particles in the condensate — that is, with minimum velocity — grows slowly as the temperature is decreased until, just at absolute zero, they are all in this lowest energy state.

What's more, London suggested that a macroscopic current could exist in helium II; this current would represent the collective motion of the entire condensed liquid. In other words, for the second time London was proposing an explanation of a macroscopic low-temperature phenomenon as a quantum effect. For superconductors the macroscopic current is the one that creates a magnetic field of the same size but opposite in direction to the external field. For superfluid helium his idea was still vague, but, as he emphasized with this type of model: "An understanding of a great number of the most striking peculiarities of liquid helium can be achieved without entering into any discussion of details of molecular mechanics, merely on the hypothesis that some of the general features of the degenerate ideal Bose-Einstein gas remain intact, at least qualitatively, for this liquid." [5]

London was still looking for a stable job at this point, around 1937. While he was in Paris, he had the occasion to discuss his ideas with another refugee, Lazlo Tisza, who was himself in a precarious situation. To London's great annoyance, Tisza literally took hold of his suggestions and used them to develop a model that we are already familiar with — the two-fluid model. In superconductors one of the fluids consists of the electrons that have "normal" behavior and, the other, the "superconducting" electrons.

Tisza's idea of the two fluids was quite simple. One component was a superfluid, the condensate, with zero viscosity and density ρ_s. The other was a normal fluid with density ρ_n and a viscosity close to that of helium I. The density of liquid helium was the sum of the two. At absolute zero the entire liquid was supposed to be superfluid, so ρ_n was zero. Now we are close to the two-fluid model of superconductivity of Gorter and Casimir, in which one component had no electrical resistivity and the other had normal electrical properties. What was different for helium was that the superfluid component behaved like a Bose-Einstein gas. The paradox of the apparent discrepancies between the results of the different types of viscosity measurements could now be explained quite simply. The experiments with the oscillating disks measured only the normal component of the viscosity — the superfluid component did nothing to stop the disks, but they stopped anyway because of the normal component. On the other hand, the experiments with the capillary tubes measured only the superfluid component; the normal component could not penetrate extremely narrow tubes because of surface tension.

The article by Tisza immediately sparked new work by Fritz London and, independently, by his brother Heinz as well as further calculations by Tisza himself. [6] In particular, Tisza was able to predict that the two components could be spontaneously moving relative to each other in such a way that the density of each component oscillated while maintaining the overall density constant. (No matter how strangely helium behaves, it must still obey the law of conservation of mass.) Because the density of each component is related to its temperature, this exchange between the two phases could lead to temperature oscillations, which Kapitza discovered shortly thereafter.

Everything seemed to fit. The theory explained an experimental paradox and had some predictive power. There was, however, a big problem: Tisza's model was based on a physical impossibility. How could one distinguish between two fluids made of the same atoms at the same temperature and pressure, fluids that should be indistinguishable? Let's be honest: we could have made the same remark about the two superconducting fluids in chapter 5. The fact that it is only now that we bring up this issue is because it became a driving force in theoretical research on helium. The model of Tisza and London had to be modified. It could no longer be based on distinguishing between indistinguishable atoms, but it still had to explain the phenomena associated with helium II.

The program was clear, but the goal was achieved only at the price of lively polemics. London was aware of the deficiencies of the model, but so were Landau, just out of prison, and Kapitza. Research on helium practically stopped in Western Europe in 1938; in the Soviet Union, however, such research continued a while longer because of the respite furnished by the Soviet-German treaty. In fact, work continued right up to the German invasion of the USSR, since the Institute of Physical Problems that Kapitza had created didn't move to Kazan, a city about a thousand kilometers southeast of Moscow on the Volga River, until 1941. It was just in that year that Kapitza summarized his work on helium in an article in the Soviet publication the *Journal of Theoretical and Experimental Physics* (*JETP*),[7] and Landau simultaneously presented a theory of superfluidity that at first sight was radically different from the ideas of London and Tisza.

Because they were dealing with a Bose-Einstein gas, Tisza and London treated the wave function of each atom individually, as if it moved about independently of all the others, like molecules of air in a room. Landau thought this was nonsense. Helium is a liquid, he emphasized, so the atoms exert strong forces on one another and could not be considered as independent of one another. He therefore turned his back on this microscopic approach and returned to the classical equations of hydrodynamics. Assuming that the physical quantities in these equations could be "quantized" — that is, transformed into quantum mechanical operators — he changed the classical equations into quantum equations. In other words, Landau treated this liquid the way a contemporary physicist treats a solid. To calculate its specific heat, he thinks not about the change in the motion of each atom with increasing temperature but, instead, about the change in the motion of the solid as a whole, its "excitations." Thus, Landau insisted that there were not two types of atoms in the helium but, rather, two types of motion. He made his point of view clear in a review article:

> When we consider helium as a mixture of two liquids, we are merely having recourse to words to describe what's happening in helium II. Like every attempt to describe quantum phenomena in classical terms, it is not really adequate. The proper way to say it is that in a quantum liquid two kinds of motion can exist simultaneously, each with its own "effective mass," such that their sum is the actual mass of the liquid. One of these is normal, i.e., it follows the usual laws for ordinary liquids; the other is superfluid. These two kinds of motion take place without any transfer of momentum between them [i.e., no force of one on the other]. We insist on the fact that there is no separation of the real particles of the fluid between those that are superfluid and those that are normal. In a certain sense, one can speak of "superfluid" masses and normal masses of the fluid since each of the particles is connected with one of the two possible types of motion. But in no way does this mean that the real liquid can be physically divided into two parts.[8]

The tone is quiet and conciliatory, yet Landau gives his opponents no quarter: "The explanation put forward by Tisza has no basis whatsoever in his own suggestions, but is, in fact, in direct contradiction with them." This sharp comment reflects the tone of the combative letters that the two exchanged during the entire war. Ginzburg, Landau's inseparable colleague, followed the same path with even less nuance: "Bose condensation of a gas has absolutely nothing to do with the properties of helium. Only Landau's theory explains them."

One of the Soviet experimentalists, V. P. Peshkov, a specialist in helium, rejected the London-Tisza theory as "very artificial and hardly convincing." When he found out about this comment, London replied: "In any case, the Bose-Einstein gas has furnished an idealized model that is valid for almost all properties of liquid helium, a model that, in addition, has led to the prediction of new effects. . . . Landau's theory arrived too late for us to treat deductions from it as real predictions." Tisza himself did not stand back from the fray: "One must prefer Bose-Einstein theory to the quantum-hydrodynamical approach."[9] In fact, the two approaches turn out to be equivalent, provided that the particles of the Bose-Einstein gas are not considered to be independent noninteracting helium atoms. As shown by N. Bogoliubov in an article that appeared in Russian in 1947, of which Tisza was obviously not aware, liquid helium can be considered as a Bose system modified by interactions between the atoms.[10]

It's easy to understand that Tisza knew nothing of Bogoliubov's paper. What is more surprising is that his Soviet colleagues did not appreciate Bogoliubov's contribution. In 1946 the word spread among the young physicists in Moscow that a newcomer had arrived from Kiev loaded with talent and full of imagination: Nicolas Nikolaievich Bogoliubov. Naturally, like Tamm, Ginzburg, and many others, he worked on the Soviet nuclear program, trying to catch up with the Americans,[11] but, exploding with ideas, he also became interested in helium. Andrey Sakharov was in the audience for a seminar he gave to present his results: "I attended a remarkable lecture by Bogulioubov on superfluidity [at the Institute of Physics]. Of course, it was just a model theory, and, moreover, one that relied on perturbation theory, but it was the first theoretical investigation that derived the surprising phenomenon of superfluidity from first principles without relying on a specially postulated spectrum of elementary excitations."[12]

The a priori choice of a spectrum that gives just the right answer is evidently Landau's, chosen with his remarkable insight. Bogoliubov had adopted a radically different approach, trying to find a fundamental mechanism based on boson formation (particles with integer spins) that would justify the chosen spectrum of excitations. Forty years later, in his exile at Gorky, Andrey Sakharov described what followed: "Unfortunately, certain scientists did not appreciate his work, and Bogulioubov, to say nothing of his students and associates, engaged in some rather dubious conduct during the squabbles that followed. Ten years later, however, when articles on superconductivity by Bardeen, Cooper, and Schrieffer appeared, Bogoliubov had a suitable theoretical framework ready, and he capitalized on it brilliantly."[13]

In fact, it appears that some people around Bogoliubov whom we know little about were prejudiced against Landau's work (with grave consequences for Bogoliubov's future). It is important to note, however, that Landau's school did not seize upon the fact that Bogoliubov's ideas held the key to a microscopic understanding of Landau's brilliant phenomenological insights. By the same token, Bogoliubov did not refute Landau. The same physical phenomenon can be described in different ways, and the relations between these different theories can be complicated. Physical theories can be embedded inside one another like Russian dolls, one theory being a special case of a more general theory. Bogolioubov was undeniably the first to provide a microscopic description of a quantum fluid. That does not mean that he invalidated the phenomenological approach, not at all.

London had reproached Landau for not making any predictions about superfluid helium. Little by little, however, Landau's idea of elementary excitations became a common thread in his papers, and this idea was later widely used in condensed-matter physics.

The naive image that we have from classical mechanics is that atoms fall into an eternal sleep as an object gets cold; their energy decreases to zero. At temperatures near absolute zero, the atoms oscillate only very weakly around their equilibrium position. That's why, according to classical nineteenth-century theories, all bodies solidify at low temperatures and form a crystal lattice. Liquid helium is the only system in which quantum effects appear before the liquid becomes a solid. Helium is exceptional because helium atoms are so light and because they exert such weak forces on one another. In every other material the interactions between the atoms are strong enough that the interactions dominate before quantum effects can appear.

A solid at absolute zero is in its lowest energy state, called the *ground state*. Even then, there is some movement; quantum mechanics requires it. At temperatures above zero, the solid is in an excited state, with more energetic oscillations. The energies of the possible excited states of the solid do not vary continuously—there are gaps between the observed energies. The mathematicians say that they form a discrete set, like the whole numbers 1, 2, 3. . . . Thus we say that the energy of the solid is *quantized*, and each quantum of vibrational energy is a *phonon*. A phonon is not associated with the motion of a single atom. As we have seen in metals, it is a quantum of the oscillation of the entire crystal, like a quantized sound wave. Each phonon is therefore an elementary excitation of the solid and propagates within it. It has a speed of propagation and energy, exactly like a photon of light. It is a "quasiparticle."

The notion of a *quasiparticle* (for which, in part, Landau received his Nobel Prize) is a generalization of the concept of elementary excitations extended to em-

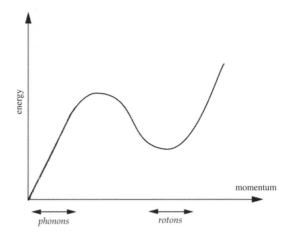

The excitation spectrum of superfluid helium. To make a liquid vibrate or rotate, energy must be provided, which is said to "excite" the liquid. Landau proposed describing the possible excitations of the liquid, the excited states, as *quasi-particles* with energy and momentum. The relationship between their energy and momentum that he suggested is plotted here; this is an excitation *spectrum* of superfluid helium. The advantage of this spectrum is that it doesn't allow very low energy excitations — by collisions with the walls of the container, for example. The liquid thus will not "feel" the rough edges of its container; it will be superfluid. Experiments have amply confirmed the shape of this spectrum.

brace an entire collection of interacting particles, whether it be an electron gas in a metal or atoms of liquid helium. You can get some idea of what this means by considering a puddle of water that is (almost) perfectly still. That's our ground state. If it starts to rain very lightly, each drop of rain will initiate a pulse of circular ripples that will expand and finally disappear if the puddle is large enough. Although each pulse that propagates in the puddle causes a huge number of water molecules to start vibrating, the pulse itself behaves like an autonomous entity that can be described by just a few parameters, such as its speed and effective mass, which are characteristic of the surface of the puddle. It is clear that this description is reasonable only if it's a light rain, so that each pulse can be described separately, neglecting any interaction with other pulses.

At the microscopic level Landau's insight was that he realized that these elementary excitations were truly quantum states to which one could attribute quantum numbers just as for particles, and so he called them *quasiparticles*. Just like the water waves, the quasiparticles exist only as long as they are few in number and interact only weakly with one another. The notion of an elementary excitation as a distinct entity makes sense only if its interaction energy is small compared to its own energy. When these conditions are fulfilled, these low-density excitations form a kind of ideal gas. In this limit, in which each quasiparticle has a well-defined energy and speed, one can define a relation for a particular material that shows how the energy of the quasiparticle depends on its speed. The graph of this relation, characteristic of the solid, is called the *excitation spectrum* of the solid.

Landau applied these ideas about ground states, excited states, and excitation spectra to helium. The phenomenological two-fluid model of Tisza seemed to give a qualitative description of the behavior of superfluid helium that worked well both for its superfluid properties and for their temperature dependence. A microscopic model could provide a firm basis for this model: it made the two fluids real. It seemed reasonable to interpret the superfluid part as a manifestation of Bose-Einstein condensation of interacting particles, in accordance with London's prediction. As for the normal component, with finite viscosity, Landau suggested that it was formed of quasiparticles, which represented the excitations of the system. He had the extraordinary intuition of introducing not one but two types of quasiparticles. To the usual phonons he added a second type of elementary excitation that, on the suggestion of his colleague Tamm, he called *rotons*. Tamm was another highly talented Soviet theorist; one of his students later acquired a worldwide reputation well beyond the circle of physicists. This was Andrey Sakharov.

The name *roton* itself had no particular meaning; even though it seems to suggest something about rotations, physicists did not have a clear idea of what a roton was.[14] Above all, the name did not magically provide some deep understanding of the nature of superfluidity—the physics is not in the name. Landau suggested a spectrum of helium; it required another kind of quasiparticle, and he had to name it. This spectrum is the sum of the spectra for the two types of quasiparticles. The low-energy excitations are exclusively phonons; they dominate the behavior of the superfluid at very low temperatures, up to, say, several tens of kelvins. At higher temperatures the rotons start to appear; their contribution to the energy of the system comes from the very surprising form of their spectrum, a form that Landau intuited and which was perfectly verified many years later in neutron diffraction experiments.[15]

The gas of ordinary particles in the normal fluid component of London's and Tisza's helium became a gas of quasiparticles in Landau's model, a gas of elementary excitations. The polemics between the proponents of the two models came to an end. (As it happened, even though London visited Leningrad and traveled throughout the Soviet Union down to Odessa in 1931, it seems that he and Landau never had the chance to meet each other.) These notions of quasiparticles and the like that came from liquid helium research seem nowadays to be totally ordinary for condensed-matter physicists. As a result, they tend to think that the success in explaining helium must soon have flowed over into superconductivity.

Nothing of the sort happened. The war was one reason. Also to blame was the fact that experimental methods in the two areas, liquid helium and superconductivity, were quite different. We must remember also that what is taught now right from the beginning of condensed-matter physics courses required years of work by many physicists, including Landau. Here, as elsewhere, his writing style and the abstract nature of his work made his results less than accessible. Landau was also sufficiently elitist that he made little effort to present his ideas clearly in his original articles.

When Landau proposed rotons, he was guided only by a rigorous critique of the approach of London and Tisza. The spectrum of excitations that he proposed was not based on some naive scheme as is often the case. His genius lay in his choice of the form of the spectrum. Landau knew by heart all sorts of solutions of differential equations and could immediately visualize a plot of these solutions. Like all physicists, Landau had recourse to mental images to get hold of a problem and discuss it with his colleagues, but he didn't waste time with naive representations, like the spiral trajectories of electrons that Einstein used in the 1920s to discuss superconductivity. Instead, Landau had recourse to his sophisticated mathematical culture in proposing the rather singular shape of the spectrum of helium.

London's attitude, on the other hand, is a good illustration of the difficulties that even highly talented physicists can encounter in trying to make progress in understanding phenomena that have little basis in ordinary experience or classical concepts. London, the first to put forward the idea of macroscopic quantum order, nonetheless developed a microscopic solution for helium, involving a Bose condensation of the helium atoms. Ironically, it was Landau who at this stage prolonged the macroscopic view that blinded him to the benefits of London's approach. With the passage of time the conflict between the two men and the bitterness of their polemics seem excessive. The precariousness of both of their situations and Landau's personality can perhaps explain their rancor.

True understanding of superfluidity came shortly after the end of World War II. The theory of superconductivity was developed almost a decade later. The war had a major impact on low-temperature physics and on physics and science in general.

CHAPTER 11

The War, the Bomb, and the Cold

Low-temperature physics already had some experience with the effects of war during World War I. Research was interrupted, but, once peace returned in 1918, scientists picked up where they had left off several years earlier and continued their investigations. In Leiden, with a gift of helium from the United States, it was back to the rhythm established by Kamerlingh Onnes. A world had collapsed, but Leiden had not.

World War II was different: if 1914–1918 was the chemists' war, 1939–1945 was the physicists'. The war entangled low-temperature physics in the orbit of "*big science*," of big budgets and big equipment. Nevertheless, low-temperature research came to a halt as the battlefield touched every center of cryogenics, one by one. This time Leiden was no exception.

From May 10, 1940, the Netherlands was an occupied country. The German invaders installed a government of collaborators, and Leiden became the first laboratory to suffer the repercussions of the war. On November 26, 1940, all the Jewish professors in Dutch universities were dismissed from their positions, among them the most famous lawyer in the Netherlands, E. M. Meijers, a professor in Leiden. Faithful to a long tradition of free speech going all the way back to Spinoza, the dean of the law faculty, R. P. Cleveringa, protested publicly against his expulsion, and, strengthened by this support, the Leiden students went on strike. Punishment was quick; the dean was arrested and the university closed for the duration of the occupation.[1] The laboratory was not directly affected, but the circumstances nevertheless put an end to Leiden's leading role in low-temperature physics.

The Soviet laboratories were able to pursue their normal activities until 1941, thanks to the German-Russian treaty. The purges and the departure of Landau for Moscow in February 1937 had practically stopped all work in Kharkov, so there was essentially no longer any superconductivity research in Ukraine. In Moscow, on the other hand, the study of superfluid helium, as much theoretical as experimental, continued apace and, in spite of the situation, led to the crucial contributions mentioned in the last chapter. At the start of the German invasion, the Institute of Physical Problems moved to Kazan. Priority then went to military problems, and both

Landau and Kapitza were involved. Because of the extremely cold weather in Kazan, the buildings that housed the scientists were barely habitable, and the seminars were held in bathrooms.[2]

The laboratories of the Kaiser Wilhelm Institute in Berlin-Dahlem, Germany, functioned until the first series of intense air raids that devastated Berlin at the end of August 1943, when they were almost entirely evacuated. Meissner's laboratory in Berlin had been transferred to Munich in 1940 because Heisenberg wanted to use the Berlin site for the German nuclear program that he was directing. Until the end of 1943 *Zeitschrift für Physik* published articles on superconductivity, including a series of theoretical review articles by Max von Laue.[3] These papers did not add much to the work of the Londons, but public and written praise of the applications of quantum mechanics under the Nazis was typical of von Laue.

Zeitschrift für Physik also published an article at this time by E. Justi that has been, perhaps unjustly, forgotten.[4] In fact, his article was the first in a series that seems endless, papers claiming such extraordinarily high temperatures for the on-set of superconductivity that the reader suspects a hoax. Following the tradition of Meissner, Justi was studying niobium compounds and thought he saw "seeds" of superconductivity in niobium nitride (NbN) up to 112K. In reality the transition temperature in this compound is more modestly situated in the range between 6K and 15K, depending on its crystal structure. More than forty years later superconductivity was actually observed at such temperatures, as we shall see, but in quite different materials.

Justi's work attracted no particular attention because, however unexpected its result, it didn't collide with any prevailing theory that it might have overturned. In 1943, after all, no one had any serious ideas about how to explain superconductivity and its critical temperature. This attitude was underscored when, just after the war, Mendelssohn, citing Justi's article in a review of research during the war, treated the result as a curiosity and paid little attention to it. At the end of the 1940s low-temperature physicists were not fixed on the goal of attaining higher and higher transition temperatures for superconductors. Understanding the unexplained phenomena for their own sake seemed to them much more important.

In contrast with nuclear physics and electronics, low-temperature physics had few specific benefits from the war, although it profited from the increasing importance of physics as a whole in society. Even before the closings and forced transfers of cryogenics laboratories, military research, especially that focused on the atomic bomb, had higher priority than all programs of fundamental research. Low-temperature physics had no special ranking. The same Justi whom we have just retrieved from oblivion was set to work on heat transfer studies that eventually led to a small heavy-water reactor.[5] Cryogenics was supported only when it was useful in the bomb program, for example, when the Germans briefly thought about trying to extract deuterium from industrial liquid hydrogen. Sabotage by the English and Norwegians had prevented the Germans from obtaining heavy water from the Norwegian electrolysis factory, Norsk-Hydro, the sole producer in the world. Thus, the German firm Linde, which had furnished the nitrogen liquefaction equipment for Kharkov, was commissioned to try something similar for deuterium.

Most of the time, however, low-temperature specialists were mobilized for general scientific support of the war. They had specific technical abilities that were in demand — vacuum techniques and fine soldering, for example, if they were experimentalists, computational skills and quantum mechanics if they were theorists. In fact, the reputation of the low-temperature experimentalists in gas handling and vacuum techniques, which were essential in separating the isotopes of uranium, led the Allies to overestimate the progress of the Soviet and German nuclear programs. Skilled scientific manpower was rare. In each country working on the bomb, some of the most talented low-temperature and condensed-matter physicists became involved, out of conviction, obligation, or opportunism.

Beginning in December 1940, Franz Simon led a British team that set up plans for a plant to produce uranium hexafluoride on an industrial scale by gaseous diffusion through porous membranes. His old student Nicolas Kurti was one of his first hires. Their goal was nothing less than the production of a kilogram of 99 percent pure U per day.[6] Simon, who had established his reputation with portable minirefrigerators for helium, first in Berlin and then in Breslau and Oxford, thus found himself constructing a factory that was supposed to occupy about fifty acres and consume about sixty thousand kilowatts of power.[7] When hostilities began, Heinz London was imprisoned as a citizen of an enemy country. Once released, in 1940, he was able to work until the very end of the war in Birmingham and then at Harwell alongside Simon, who had been his thesis advisor.[8]

In Germany as much as in the United States, helium played an indirect role in the nuclear program. The Germans, for example, were interested in monazite, the mineral from which Kamerlingh Onnes had retrieved his helium. Monazite also contains thorium, a possible bomb material. This caused a great deal of anguish for Alsos, a joint military-scientific team launched in the wake of the debarking Allied troops to investigate the state of the German bomb program. Alsos was led by Samuel Goudsmit, a well-known physicist who had been a student in Leiden.[9] They set out expecting the worst. Based on the history of the last quarter-century, the members of Alsos shared the general opinion that German science was the best in the world. This new breed of spy also knew that Heisenberg had stayed on in Berlin, in spite of the bombardments, to lead a series of nuclear measurements under cover of secret biological research in a "virus building" that was an air raid shelter.[10]

Nothing could have confirmed the team's pessimistic expectations more than their discovery, upon their arrival in Paris, that the monazite stored in France had just been hijacked by the Germans. For Alsos, of course, the concern over monazite was not the helium but, rather, the thorium. They knew that thorium could be useful in the fabrication of a bomb, because bombarding it with neutrons would yield fissile uranium.[11] Well in advance of the nuclear program, however, the thorium in the monazite had made it the object of a small but flourishing commerce. The German firm Auer, well known to housewives between the wars for its gas stoves and lighters, imported the raw material. Auer wanted the thorium oxide in the monazite — everyone used it to make brighter gas lamps. At the same time, like the Dutch in the time of Kamerlingh Onnes, the Germans had continued to import monazite from India and Latin America for its helium, since they lacked a supply of natural gas. But thorium was the essential commercial ingredient.

Alsos found out in Paris that before the war the French Society for Rare Earths, a French company, had a monopoly on thorium for France. Its owner was a French Jew, so the society was taken over by Auer, which transferred the entire stock of thorium to Germany just before the liberation of France in 1944. What was even more disquieting to Alsos was that the Auer chemist who arranged the transfer was the same person responsible for providing uranium to the German government. The members of Alsos were appalled: the amount of thorium that had been transferred would cover the ordinary industrial needs for thorium for twenty years. Under pressure from Washington, the Alsos spies discovered after many missteps the real reason for the theft of so much thorium: when the war was over, Auer planned to make thorium toothpaste! Doramad was a popular toothpaste in Germany before the war; it was considered to be a tonic because it contained radioactive thorium. "Your teeth will shine with radioactive brilliance," it advertised.[12] Auer had run off with the stockpile of thorium to ensure a monopoly on the toothpaste. The story seemed unbelievable. Goudsmit became convinced of its accuracy only when he questioned the first German nuclear physicists captured in Strasbourg and realized that the German bomb project was hardly getting started, even though Heisenberg was in charge.[13] Monazite was thus not about to become for the Nazis the strategic equivalent of heavy water. Today monazite has pretty much lost all its interest, even for cryogenics. It is no longer used to obtain helium, which is now recovered from gaseous pockets found at the top of oil deposits.

Cryogenics and nuclear physics also crossed paths on the American side. At an early stage in the bomb project, liquid helium was considered as possibly useful in making a nuclear reactor. To obtain a chain reaction, Fermi and Szilard had first of all thought of using water to slow the neutrons, because it contained hydrogen. G. Placzek, who had worked in Copenhagen, suggested to them in the spring of 1942 that helium be used instead. To be sufficiently dense, it would have to be liquid, and so the entire reactor would have to be cooled to 4.2K — the suggestion would not exactly ease the task of quickly and cheaply proving the principle of the reactor. For an experimenter like Fermi, it appeared absolutely ridiculous and worthy only of a theorist. Szilard said that from then on, whenever Fermi talked about helium, he always called it *Placzek's helium*.[14]

The career of liquid helium in the origins of the American program for building an atomic bomb was thus quite brief. Soon, however, the Manhattan Project in Los Alamos, New Mexico, did call for cryogenics on a grand scale. The person who wanted to bring in low temperatures to the bomb project knew absolutely nothing about how to apply them. This was Edward Teller, who had entered the history of the U.S. nuclear program by the side door: Teller had served as Szilard's chauffeur when Szilard went in search of Einstein on Long Island to write the famous letter to Roosevelt. As a friend of Fermi and Szilard, he rejoined them during the summer of 1941 when they worked together at Columbia University. On a walk together, Fermi wondered out loud whether an atomic bomb could be a strong enough heat source to initiate a fusion reaction between deuterium nuclei. If such a reaction was possible, then one could bring four hydrogen atoms together to form helium and make a fusion bomb. The energy liberated would be three orders of magnitude larger than the energy of the fission bomb, which, of course, still didn't exist.

Teller found in this problem enough to keep him doing physics for a long time. At first sight the fusion bomb did not seem feasible, and Fermi, to whom he revealed his result, was happy. A few months later Teller, having just arrived at the Metallurgy Laboratory of the University of Chicago where Szilard was working and having no specific duties, got to work with a colleague on the calculation of the possible fusion and became convinced that, on the contrary, it was indeed possible. He became the indefatigable advocate of this superbomb, especially in the summer of 1942, during a meeting of nuclear theorists organized by J. Robert Oppenheimer in Berkeley, California. The decision to construct the laboratory in Los Alamos had not yet been taken; the hydrogen bomb would not be developed until several years afterward. Nevertheless, when it came time to decide on the necessary infrastructure for Los Alamos, Teller noted that the superbomb required liquid deuterium. Among the six laboratories created, one of them was the cryogenics lab necessary to get to 17K, the temperature at which deuterium becomes a liquid under atmospheric pressure.

The history of physics was indelibly marked by World War II, primarily because of the use of the fission bomb. During the 1920s and 1930s intellectuals and sometimes the general public became used to the idea that modern physics would question concepts of reality that seemed obvious by common sense. What they didn't realize was how profoundly modern physics would affect society itself. After all, wasn't the Newtonian revolution confined to the realm of ideas? Why would the new physics be different? Such insouciance was no longer tenable after 1945, and not only because of Hiroshima.

Even if the bomb had never been created, the coupling of highly innovative physics ideas with a quasi-industrial organization of scientific work would have produced "big science." Kamerlingh Onnes started the movement in cryogenics. During the 1930s other scientific entrepreneurs appeared, including E. O. Lawrence in Berkeley and Kapitza in Moscow. Frederic Joliot-Curie in Paris was a left-wing French version of the same phenomenon. They were the pioneers. After the war no branch of physics would escape this evolution that took American laboratories as its model. Low temperatures and superconductivity were no exceptions.

Los Alamos certainly made no contribution to low-temperature physics, yet low-temperature techniques benefited from links with the enormous military-scientific program for the atomic bomb. Something similar happened in the 1960s in space exploration, which also relies on cryogenics. Once again, it is difficult to find in space programs a specific contribution to our knowledge of low-temperature physics, but the development of low-temperature techniques was supported by big science.

The new context inspired changes that were not at all limited to the social and economic organization of scientific work. The way new ideas were formed and new results appeared in superconductivity just after the war show this clearly. Without big science, developments would have occurred in a different order, and it is hardly likely that progress would have been so rapid — it would have been unthinkable before the war. Big science has also had an impact on the epistemology of science, and, in spite of what we've just said, not all its consequences have been positive.

For experimenters the transformation was radical. The amount of equipment available on the market would have been beyond fantasy before the war. Emmanuel Piore, a scientific leader at IBM and an advisor to the U.S. government at the time, once remarked that interest in low-temperature physics had been diminished by the difficulties of constructing and commissioning helium liquefiers. With no justification other than this, he convinced the U.S. government to begin supporting industrial fabrication of the liquefiers. The consulting firm of Arthur D. Little won the government contract and effectively commercialized helium liquefiers in the United States.[15] A new world was being constructed for science, and it contributed to a significant increase in the distance between experimenters and theorists. While it was essentially the great prewar figures who continued to hold center stage among the theorists, experimentalists were at the same time both more numerous and generally younger. In comparison with research on helium, superconductivity research provided easier access for scientists leaving military service. Most of the experimental techniques were extrapolations to the low-temperature domain of methods used in electromagnetic measurements that were being widely pursued. Helium research, on the other hand, required starting from zero at every step of the experimental program.

This new world of physics was not entirely dominated by nuclear physics, however, as we tend to believe with the passage of time. Thousands of young people had been engaged in scientific programs for the military, only a minority of them concerned with the bomb. But, whether they had been involved in the Manhattan Project or other areas of the scientific war effort, few of these new physicists had a broad vision of the range of open questions in physics or, even less, in superconductivity, which was not at all fashionable. A popular anecdote of uncertain source is telling. A young scientist is asked to define *physics*. "A branch of radar," he quickly replied. Radar cannot even be called a branch of physics, but it became important to superconductivity.

CHAPTER 12

Radar and Superconductivity

When the English wanted to arrange a scientific collaboration with the Americans during the war, they first communicated some results about the bomb (obtained by R. Peierls, H. London, N. Kurti, and F. Simon, low-temperature physicists all). Then they revealed a device the Americans had never seen and had barely imagined. This was a magnetron, an electronic tube for generating electromagnetic waves with frequencies much higher than radio waves, the forerunner of the tubes now used in microwave ovens. The magnetron made radar possible, a development that proved vital during the rest of the war.[1] Once peace returned, this high-frequency equipment could be found in U.S. military stockpiles everywhere. (These stockpiles enabled the French physicist Yves Rocard to relieve the penury of scientific apparatus in France. He scrounged from secondhand dealers in New York to outfit the physics laboratories of the Ecole Normale Supérieure in Paris.) High frequencies had already played a role in the history of superconductivity; as we noted earlier, Heinz London was studying the high frequency resistance of a superconductor when he and his brother began to attack the problem of the electromagnetic behavior of superconductors.[2] Heinz London had given up superconductivity, however, and was primarily involved in the British nuclear program. Yet high frequencies would turn out to be important in confirming an important aspect of the London model.

How could one measure the London penetration depth, the distance over which the magnetic field decreases significantly as it tries to penetrate the surface of a superconductor? The basic method is simple, in principle. Take a superconducting sample with a volume that can be easily determined by measuring its geometric dimensions. Then measure the volume from which the magnetic field is excluded. Subtraction yields the magnetized part, and, with a bit of straightforward manipulation, the penetration depth can be deduced. A fluxmeter is needed to determine the volume with no magnetic field, that is, the region where the Meissner effect is perfect. A fluxmeter is a coil wrapped around the sample and connected to a galvanometer; it measures the amplitude of the signal produced when the flux is expelled from the superconductor. The size of this signal depends on the product of the applied magnetic field and the field-free volume.

The experiment just described cannot be done; a fruitless attempt by Casimir demonstrated this amply. London had estimated that the penetration depth was very small, of the order of a tenth of a micron. With an ordinary sample a few centimeters on a side, the subtraction involves two large numbers, each at least somewhat uncertain because measurements are never perfectly precise. An effect of the size London had estimated would then yield a tiny difference in volume with a very large uncertainty.[3] This doesn't mean that Casimir's experiment was poorly motivated — London might have been wrong, as theorists often are. Only a sample with dimensions of the order of the predicted penetration depth, however, would be appropriate for measuring such a small effect. Then the problem would shift to the fluxmeters — before the 1960s, at least, they weren't very sensitive. To get a large enough effect, a large number of small samples would have to be measured at the same time: the sample had to be a superconducting powder. As Fritz London described it after the war: "In most experiments, the supercurrents can be simply considered as surface currents with no thickness at all. The determination of the penetration depth in macroscopic superconductors requires precision measurements. It would therefore appear preferable to use very small superconductors comparable in size with λ in order to obtain big effects."[4]

David Shoenberg applied this idea slightly differently in 1940 by measuring the susceptibility of a highly dilute suspension of superconducting grains. With a colloidal suspension of microscopic beads of mercury,[5] he became the first to demonstrate the variation of the penetration depth with temperature.[6] Shoenberg was able to show something that London had not thought about, that the penetration depth becomes larger and larger as the temperature increases toward the critical temperature. This was easy to understand in the two-fluid model of Casimir and Gorter. As the temperature increases, the fraction of superconducting electrons becomes smaller, they become less efficient in attenuating the magnetic field, and the distance over which the field penetrates becomes longer and longer.

These experiments by Shoenberg in Cambridge were based on the hypothesis that microscopic superconducting samples behaved like macroscopic samples of ordinary size. No one knew whether this scaling hypothesis was correct. Resolving this problem returns us to the postwar period and applications of the high-frequency equipment developed during the war. In spite of the difficulties of everyday life in England, the Royal Society Mond Laboratory at Cambridge was the obvious site for beginning once more the exploration of the penetration depth. The intellectual tradition, the talent, and the equipment were all united under this one roof. In 1947 the rich supply of superb high-frequency equipment allowed a young and elegant physicist, Brian Pippard, to perfect a new method of high-frequency measurements to determine the penetration depth.

With direct currents the electrons are spread uniformly over the cross-sectional area of a conductor, because the electric field is the same everywhere in the conductor. This is not true for an alternating current, in which the direction of electron flow switches back and forth. In this case, as noted previously, the current is concentrated near the surface of the conductor in a "skin" whose thickness decreases as the frequency increases. More precisely, the thickness is inversely proportional to the square root of the product of the frequency of the current and the conductivity of the metal. The effect is aptly named the *skin effect*. At high frequencies, beyond 10^9 hertz, the skin is about one micron thick. In other words, at such frequencies, the penetration depth λ (estimated by London to be about 0.1 micron) is not much smaller than the skin in which the current exists when the sample is resistive. It is thus possible to measure λ by comparing it with the skin rather than the whole thickness of the superconductor. This is the reasoning Heinz London had already used in 1934. At that time, however, he had no way to do the measurement, as was also true several years later when two Soviet physicists tried it.[7]

Pippard, with all the technology of radar at his disposal, constructed a cavity with a resonance frequency of 1200 megahertz (1.2×10^9 hertz) whose end plate was superconducting.[8] By measuring the width of the resonance peak, he could deduce the London penetration depth precisely in tin and mercury. His experiments confirmed with much less uncertainty the earlier numbers obtained by Shoenberg.[9] Indeed, λ proved to be a quantity that became infinite as one approached the superconducting transition temperature from below. The value of λ also turned out to be independent of the strength of the applied magnetic field (provided the field was small enough to maintain superconductivity). Fritz London, who had always been critical of the style of working which went hand in hand with big science, nevertheless did not hide his admiration for the elegance of Pippard's experiments, which was due almost entirely to the new techniques: "The amazing development of microwave technology during World War II also brought great progress in the performance of these high frequency experiments which were previously extremely difficult to carry out."[10]

Pippard also discovered an effect that was quite unexpected at very low temperatures: the high-frequency resistance of tin in the normal, nonsuperconducting state was eight times larger than its resistance at low frequency under the same conditions. The measurements were difficult because the resistance was still small. The increase was quickly explained by G. E. H. Reuter and E. H. Sondheimer:[11] at high frequencies the skin depth becomes so small that there is little free space for the electrons to move about, so the frequency dependence changes and the measured resistance becomes higher.

What is important here for the history of superconductivity are not the details of Pippard's discovery but, rather, that, early in his career, he encountered interesting effects of the electron's *mean free path* (the average distance between collisions). When the mean free path becomes very long, longer than the high frequency skin — that is, in the "anomalous skin-effect" region [12] — the equations of electromagnetism have to take into account not only what happens at a particular point but also what happens in the entire surrounding neighborhood. In effect, the electron covers such large distances that the strict notion of "locality" no longer makes sense. The electromagnetic phenomena are then said to be "nonlocal"; they do not depend only on what happens at a particular point.

Ideas closely related to this work imbued Pippard's work on superconductivity. He now began to look at superconductivity in metals with impurities. Adding impurities to a normal metal adds obstacles to the motion of the electrons, decreasing the mean free path and increasing the resistance. For superconductors London had started with the two-fluid model of Gorter and Casimir, and he was interested only in the superconducting electrons, that is, those that encounter no obstacles. In this approach changes in the mean free path of the normal electrons shouldn't matter at all. In fact, London's expression (formula) for the penetration depth depended only on the mass, the charge, and the density of the superconducting electrons.[13]

What did Pippard find? In adding only 3 percent indium to his tin samples, he obtained an alloy with a penetration depth twice as large as that for the tin alone. What makes this even more surprising is that indium and tin are superconductors with similar critical temperatures and magnetic fields. Pippard came to the conclusion that "the behavior of a superconductor is in some manner controlled by an electronic interaction of rather long range," which he called the *coherence range*.[14] The idea was that this coherence range made it impossible for the electrons to change their properties too abruptly from one point to another in the superconductor. Pippard put forward this idea quite tentatively, but his experiments with impure superconductors pushed him to raise questions about London's formulation. Tin-indium alloys were easy to make because both metals have very low melting points, 230°C for tin and 160°C for indium. Systematic study showed that the penetration depth in the alloys increased as the mean free path of the electrons became smaller. In addition, the change became large only when the mean free path was comparable in size to the penetration depth in the pure metal. These ideas and results for the penetration depth and the mean free path in impure superconductors began to sound analogous to the steps that led to the discovery of the anomalous skin regime in very pure normal metals that are not superconducting. Remember that a perfect conductor is not a superconductor because it lacks an essential characteristic, the Meissner effect.

The anomalous skin effect becomes significant when the thickness of the skin that attenuates the high frequencies is about the same size as the mean free path of the electrons. This problem can be resolved by noting that the value of the current at a given point depends on the electric field in the entire surrounding region. This means that the electrons arriving at a particular point have hardly scattered at all, and so they "remember" the region that they have just traversed.

For superconductors London had written an equation for the current density of superconducting electrons at a given point which depended only on what was happening right at that point; it cared not at all about what was going on in the vicinity of that point. In physics jargon it was a *local* equation (even if it yielded the same solution everywhere when the superconductor was completely homogeneous.) Pippard transformed it into a *nonlocal* equation: using an expression for the anomalous skin effect as a model, he wrote an expression for the supercurrent density at a point that depends not only on the point but also on the neighboring volume. The extent of this region was the coherence range. After many experiments Pippard generalized his coherence range to the case of superconducting metals with impurities.[15]

At first sight Pippard's equation seemed to be only a simple modification of the London equation; in reality the change was profound, even if it took some time to appreciate this. For London the wave function of the superconducting electrons was absolutely rigid in the presence of a magnetic field. It varied not at all from one point to another within the volume of the superconductor, except close to the surface, within the penetration region. Now, metals often have a mixture of superconducting and normal regions; the problems in demonstrating the Meissner effect prove this. According to London, this meant there must be many interfaces between the two kinds of regions in the metal. At each one there is a sudden transition, within one penetration length, between normal metal and superconductor. By introducing his coherence range, Pippard made this transition less brutal; it could happen only over a distance at least as long as the coherence range. In fact, in his studies of soft superconducting metals such as indium and tin, Pippard had found that, at temperatures not too close to T_c, the coherence range was about one micron, at least one order of magnitude larger than the penetration depth. This made the description of the interfaces between resistive and superconducting zones more realistic.[16]

The introduction of this second length characteristic of a given superconductor was an important contribution to the hydrodynamic description of the phenomenon. It confirmed that this approach remained fruitful even if hydrodynamics could not explain why this second length had to be introduced.

It took several years for Pippard to present his notion of a coherence range in a definitive and convincing manner. At the second International Conference on Low Temperature Physics, held in Oxford in August 1951, the idea began to take root. Herbert Fröhlich, starting from entirely different premises, talked about a spatial order parameter distinct from London's penetration length.[17] London immediately expressed his fondness for this suggestion, saying that such a length would make it possible to obtain the surface energy required to make the electrodynamics of superconductors compatible with the thermodynamics and the existence of the Meissner effect.[18]

The Cartesian spirit that has guided us up to now will note that, if superconductors are characterized by two lengths, it must be interesting to compare them. For superconductors with low melting points, the coherence range is always longer than the penetration depth of a magnetic field; these were called *Pippard*, or *type I*, superconductors. On the other hand, systematic measurements for alloys (which have a short mean free path in the normal state) showed that the penetration depth is always longer than the coherence range. These were quickly labeled *dirty*, or *type II*, superconductors.[19]

The fog lifted. This classification of superconductors into two categories based on the relation between the two characteristic lengths clarified the results of the magnetic measurements. In Pippard superconductors the magnetic field penetrates into the superconductor all at once when the field strength reaches the critical value and the metal goes normal. Dirty superconductors are different. Here the magnetic field penetrates little by little as the field strength increases, and the critical fields are much higher, so that the resistive state returns at much higher field strengths than in Pippard superconductors. Pippard suggested that this difference was the result of interface energies between the normal and superconducting regions that were positive for pure metals and negative for dirty superconductors.[20] Positive energies would tend to keep out the external field, whereas negative energies would tend to pull it in. This would explain why, for dirty superconductors, the Meissner effect is perfect only in very weak external fields.

As time went on, physicists became convinced that the coherence range, initially introduced just to improve a calculation, was in fact a fundamental quantity for superconductivity. Over this period different kinds of experimental data accumulated, not just from high-frequency experiments. Thermal measurements, the source of so much grief for the Dutch before the war, were now much easier and more precise. A sharp discontinuity in the specific heat of superconductors was discovered just at the critical temperature. As the temperature decreased toward absolute zero, the specific heat fell exponentially and not as the cube of the temperature (T^3), as previously believed.[21] The power law dependence had been easy to accept because the contribution of the crystalline network to the specific heat behaves just this way. But the exponential behavior turned out to be a crucial indicator of an energy difference between the superconducting state and the normal state, the famous *energy gap*. It neatly tied together a number of other experimental indications of such a gap, which was roughly equal in magnitude to the thermal energy at the critical temperature.

Bardeen, about whom we will hear much more, kept a sharp eye on publications about superconductivity. He soon showed that Pippard's ideas about the electromagnetism of superconductors were compatible with the existence of an energy gap. In some sense Pippard had initiated a rite of passage for theorists: from then on, no one could envision a model without a gap.

And so, because of radar, the electromagnetic approach to superconductivity found new life. Gorter and Casimir had first elaborated this point of view in their explanation of the Meissner effect, and London had taken it up again with his own interpretation. Understanding the nature of a fluid remained the central problem of superconductivity. This was thus a hydrodynamics approach; the fact that superconductors were solids remained secondary. Adopting such a framework now seemed much more justified, given the decisive progress in understanding the superfluidity of helium during this period. Given the quarrels among London, Tisza, and Landau, we cannot speak of a "school" here. Nevertheless, there is no doubt that there are similarities, epistemologically speaking, which unite Landau to London, Shoenberg, Tisza, and Pippard.

Along with radar, which nourished the hydrodynamical perspective, nuclear physics, with its isotopes, made possible another remarkable advance at this time. This progress proved decisive because it gave birth to a new approach to superconductivity, one that did not neglect the fact that a superconductor is also a solid with an ionic lattice.

CHAPTER 13

The Ions Also Move

Oak Ridge National Laboratory near Knoxville, Tennessee, put isotopes up for sale for the first time in 1948, another byproduct of the military research programs of World War II. *Isotopes* are atoms of the same element with different atomic masses; they became important tools for low-temperature physicists.

> Elements are defined by the number of protons in the nucleus of the atom. Isotopes are nuclei with the same number of protons but different numbers of neutrons. Most of the mass of an atom is in the nucleus. Thus, helium 3 (^{3}He), with one neutron less than helium 4 (^{4}He), weighs about 25 percent less than helium 4; both are isotopes of helium. Only about one atom in a million of ordinary helium gas is helium 3 — a discovery, by L. Alvarez and R. Cornog, that was not made until 1939.[1]

Helium 3 was the first isotope put on the market at Oak Ridge; because its mass is so different from the mass of helium 4, it was easy to separate the two. Helium 3 also seemed likely to be useful in research on superfluidity, which, by 1948, had become intense once more; the specialists were first in line for it at Oak Ridge. In fact, a lively debate was under way over whether Bose-Einstein statistics was essential in achieving superfluidity. Bose-Einstein statistics was required in helium 4, the one known superfluid, because it has an even number of particles. Helium 3 has an odd number of particles, so it must obey Fermi-Dirac statistics. The commercialization of helium 3 thus looked like the key to solving the problem, "a new possibility . . . for deciding, by experiment, the relevancy of the Bose-Einstein way of counting, without entering into the obscurities of complicated and approximate calculations."[2]

The reasoning of the experts appeared flawless. If Bose-Einstein statistics explained superfluidity, helium 3 should never become superfluid because it obeyed the wrong statistics. To be certain of this negative result, the experimenters would have to study helium 3 over the entire temperature range below 3.2K, where it liquefied, and, for a long time, these initial expectations were confirmed.[3] When all was said and done, however, the idea of clarifying the problems of superfluid helium 4 by studying helium 3 turned out to be rewarding, but not because it simplified the issue of helium 4. By concentrating on the light isotope of helium, low-temperature physicists founded a new branch of condensed-matter physics that has remained fertile.

In contrast to the enthusiasm that greeted the introduction of isotopes into studies of helium, in 1950 the idea of comparing the superconducting properties of two isotopes of an element seemed just silly. As H. Fröhlich reported afterward:

> It should perhaps be mentioned in this context that before 1950, every "expert" knew that the ions, being so heavy, would have nothing to do with superconductivity. In fact Kamerlingh Onnes, using mercury from meteors, seemed to have shown that the transition temperature does not depend on the isotopic composition, and a laboratory had rejected an offer from Harwell [the English nuclear center] of isotopes for the measurement of superconductive transition temperatures as a waste of time.[4]

In fact, the first experiments with isotopes in superconductivity seemed inconclusive. When a consensus was established on just what result had been obtained, however, a crucial new concept had been established—the critical temperature for superconductivity is not the same for different isotopes of the same element. This is called the *isotope effect*.

In 1986–1987, on the other hand, new superconductors with high critical temperatures were discovered, and their chemical compositions were not at all simple. This time the experts "knew" that experiments with isotopes were guaranteed to produce decisive information about the mechanism responsible for this unexpected superconductivity. Faced with a wide variety of data here at the beginning of the twenty-first century, we are still waiting for the clouds to part and reveal a clear answer. What the experts forgot or, more often, didn't know about the early experiments with isotopes was how difficult they were. Proving that the isotope effect exists was not nearly as easy to establish as a brief mention of the result in a monograph on superconductivity would lead you to believe. The fact that Kamerlingh Onnes, working with normal elements and not pure isotopes, found a negative result was not a coincidence.

The isotope effect was discovered simultaneously by two American groups, E. Maxwell at MIT, and C. A. Reynolds, B. Serin, W. H. Wright, and L. B. Nesbitt at Rutgers University. The two articles appeared in the same issue not of *Nature*, like the articles on superfluidity, but of *Physical Review*, the journal of the American Physical Society.[5] The difficulties in carrying out these measurements have been described by Bernard Serin in an article published in 1955 in the first volume of a new annual review edited by Gorter, called *Progress in Low Temperature Physics*. The first data were taken at temperatures just below the critical temperature. In this region the critical magnetic field (above which there is no longer any superconductivity) is very small and approximately proportional to the difference in temperature from the critical temperature. A signature of the disappearance of superconductivity is the penetration of the magnetic field into the sample. So, the plan of the experiment was, for each temperature, to record the value of the magnetic field that first penetrated the isotopic sample and to repeat this for a number of temperatures. Using the proportionality between critical magnetic field and temperature, this series of values of the penetrating magnetic field could be extrapolated to zero field; the corresponding temperature was the critical temperature for that isotope.

The experiment was difficult and tedious because the temperatures had to be measured precisely while working in weak (and fluctuating) magnetic fields. In addition, the first measurements were made on two isotopes of mercury with only small mass differences, 199.5 versus 203.4 units, so the difference between the critical temperatures of the two isotopes was only four-hundredths of a kelvin degree. Happily, at just about the same time but for completely different reasons, J. M. Lock, Pippard, and Shoenberg were measuring the variation with temperature of the critical field of tin. Now the mass range of the many isotopes of natural tin is almost 10 percent; it was a much better candidate for the isotope effect experiments and provided confirming results.[6] Soon the experiments were also extended to lead and thallium. The conclusion of all these experiments was that the product of the critical temperature T_c and the square root of the mass M of the isotope is approximately constant for a given element. The larger the mass of the isotope (and thus the larger the mass of the lattice ion of this isotope), the lower was the critical temperature. Since the critical temperature depended on the mass of the isotope (and ion), it meant that the experts who believed that the ions of the crystalline lattice had nothing to do with superconductivity because they were too massive were massively wrong.

Serin understood immediately that he had an important result. He called a theorist who worked a few miles away from his laboratory at Rutgers University and had just become famous with his work on the transistor, which was to revolutionize electronics.[7] This was John Bardeen, and the transistor effect brought him his first Nobel Prize. Bardeen shared Serin's enthusiasm. For the first time in years the experimentalists, who had tested all pathways imaginable, had finally found the right way: the driving mechanism for superconductivity was an interaction (a force) between the electrons and the vibrating ions of the crystal lattice.

Bardeen had been interested in superconductivity since 1935 when he discussed the work of the London brothers with his physics mentor, John Slater. Bardeen carried away one idea from these discussions: "That one should first try to understand the Meissner effect and that if one could account for this one could also account for the infinite conductivity."[8] What Bardeen called understanding the Meissner effect was quite a different matter from the explanation already provided by the Londons in the middle 1930s, which still remained the source of his inspiration throughout his work on superconductivity. According to Bardeen: "This [the

London theory] implied large orbits for the electrons, which are also required by the uncertainty principle from the fact that it was likely that only electrons with a small momentum difference from the Fermi surface are involved. Slater gave a qualitative explanation on this basis." [9]

From the moment he became interested in superconductivity, Bardeen searched for an explanation of the detailed interactions responsible. Serin's call could not have come at a better time. Just before being drafted in 1940, Bardeen had decided not to publish an article in which he tried to develop a theory of superconductivity based on the idea that electrons near the Fermi surface were affected by the displacement of the ions from their equilibrium position. He published only a summary. [10] This, moreover, was not the only problem in condensed-matter physics for which he had calculated the effects of the ion-electron interaction. It's enough to say that, as he set to work on this problem, Bardeen had every hope of resolving with one fell swoop the enigma of superconductivity.

Yet Bardeen wasn't the only physicist starting to work on superconductivity in 1950. American universities often have summer schools to which distinguished foreign visitors are invited. In addition, American physics departments were prosperous at the beginning of the 1950s, while their peers in Britain remained in postwar poverty. Even universities situated in unlikely spots in the Midwest could thus afford excellent British professors. One of them, Purdue University, an old school of aviation in the middle of an immense cornfield south of Chicago, was thus host in the spring of 1950 to a renowned theorist, H. Fröhlich. Alone in his own universe, unaware of all the latest developments, Fröhlich had just finished working on a radically new explanation of superconductivity. This theory contained the key idea that is still the basis of the modern theory of superconductivity. Fröhlich, independently of Bardeen, came to the conclusion that superconductivity occurs because of an interaction between the electrons and the ions of the crystalline lattice via the intermediary of the quantized vibrations of the crystalline lattice (the phonons).

Everyone knew that electrons repel one another strongly, because they all have the same negative charge. Physicists knew also that the effects of the Pauli exclusion principle on the electrons gave rise to what looked like another force between them, known as the exchange interaction; it was important in understanding the magnetic properties of metals. Fröhlich had a different, almost unimaginable, idea: he suggested that two electrons were, in a certain sense, attracted to each other. It happened when one electron made the ions vibrate, thus "emitting" a phonon, and the other "absorbed" this phonon. Now this mechanism was highly abstract, and its author didn't provide an intuitive physical description. It was, moreover, incomplete, but once again it suggested that the ions of the crystal lattice could play a role in superconductivity, in spite of their being much more massive than the electrons.

Fröhlich's hypothesis ran exactly contrary to the accepted wisdom that only electrons were involved in superconductivity. Having formulated the hypothesis of an interaction between electrons via the phonons, he showed that the energy of the superconducting state differs from that of the resistive state by a definite nonzero value, the energy gap. If the superconducting state is stable, it means it has a lower energy than the normal (resistive) state, and, in order for the superconductor to pass into the normal state, energy at least equal to that gap must be provided. Presuming the existence of such a gap, Fröhlich derived the correct dependence of the gap on the mass of the isotope.

Rich in the fruits of these reflections, Fröhlich left his hermitage in the Midwest in the middle of May. He set out for civilization—the East Coast and, more precisely, the sacred ground of solid-state physics, Murray Hill, a lush area of New Jersey that was home to Bell Telephone Laboratories and John Bardeen. After Serin's telephone call, Bardeen had arrived at the same conclusions as Fröhlich. So, he announced to Fröhlich, who was completely astounded, that his hypothesis about an electron-phonon interaction had just been experimentally confirmed. Fröhlich had only to add a short note to his article, which appeared at the same time as Bardeen's. These two articles outline the program needed to achieve a microscopic theory of the mechanism of superconductivity.[11] What had to be studied were the interactions between electrons and phonons: what might seem to be a problem in particle physics was, in fact, a problem in condensed-matter physics.

It is curious that the theoretical consequences of the isotope effect were only poorly understood by the experimenters and, in particular, by Serin, who was the most active among them. If an experimentalist of his quality was confused, then we shouldn't be surprised that most of the others were locked into a series of measurements with no clear line of attack on the real problem.

Serin elaborated on his point of view in his 1955 review article.[12] Rather than content himself with summarizing a number of studies that had already appeared on the subject, which would satisfy everyone but enrich no one, Serin really tried to clarify the physics. Of course, he accepted the idea of an energy gap. Like most experimentalists at the time, however, and like a small number even today, he remained attached to the two-fluid model, which was so easy to understand. He thus had to defend the idea that the two-fluid model is compatible with the electron-phonon interaction implied by the isotope effect.

In his two-fluid model Serin included an energy gap: "[The critical magnetic field at absolute zero] H_0 varies approximately as the inverse of the square root of the isotopic mass, so that the energy gap in the two-fluid model varies inversely as the isotopic mass."[13] The language in this sentence abuses two theories. For Fröhlich and Bardeen the existence of a gap in no way implies that there are two types of electrons. The two-fluid model of superconductivity had no more reality for them than it had in superfluid helium. Serin, however, was doing his best to save the two-fluid model, as we can see from the following deduction that we find a bit bizarre: Serin found the existence of a gap inevitable "because the frictionless currents [the London supercurrents] are clearly associated only with the electrons and the isotopic mass is a property only of the lattice."[14]

Serin would have us believe that the existence of a gap proves that there is a radical separation between the superconducting electrons and the phonons of the lattice. While he appears to be explaining the results of Fröhlich and Bardeen, this sentence in fact turns their idea on its head by implying that superconducting electrons do not interact with phonons. For the two theorists it was perfectly clear that, on the contrary, it is the very existence of the interaction of the electrons with the phonons that explains the gap. An incidental remark by Serin shows that he (and he was not alone) had not at all grasped this essential demarche of Fröhlich and Bardeen, who, he said, "do not demonstrate the cooperative nature of the superconducting transition."[15]

Fröhlich knew nothing about the experimental data on the isotope effect. We see here that the experimentalists had their own blind spots about theory. Thus, two new tools that came out of wartime research, radar and isotopes, led to decisive experimental results and opened up two different paths for the theorists. The one, which we label as the *hydrodynamic approach*, was based on the electrodynamic properties of superconductors; the other, the *microscopic approach*, tried to understand why the phenomenon exists by relating it to the structure of the material. Remarkable contributions were made by each side, but they were intertwined chronologically in such a way that the proponents of each approach could, in good faith, disagree with one another, so different did their starting points, their perspectives, and their theoretical methods appear.

The natural course of this debate had been greatly disrupted by the effects of the cold war since 1950. The protagonists did not even realize what was taking place.

CHAPTER 14

East Is East, and West Is West

What a year it was, 1950. Not only was it the year that the isotope effect in super-conductors was discovered. It was also remarkable for an epistemological event that we think was unprecedented in the history of science: two parallel contributions to understanding superconductivity were made that year concerning what we have called the *hydrodynamics approach*. The one, completely original, in the East, was due to Ginzburg and Landau in the Soviet Union. The other, totally independent, in the West, was made by Fritz London in the United States and was included in the first volume of his fundamental monograph.

The cold war was at its height; the hot war in Korea had just broken out. In the USSR the Lysenko affair was about to ravage Soviet biology for a half-century; in the United States the McCarthy witch-hunt was spreading like a raging river overflowing its banks. In this climate just about everyone in Western science was simply unaware of the enormous contribution of Ginzburg and Landau, while London's book with a similar outlook profoundly influenced the evolution of the field.[1] This was all the consequence of a random act of an anticommmunist union — it's not every day that longshoremen have a role in the history of physics.

Soviet journals were sent to the United States in those days by ship. The long-shoremen's union in New York City, hardly a model of democracy, decided to boy-cott all the publications coming from the Soviet Union as a contribution to the war on communism. When the ship transporting several volumes of the *Soviet Journal of Experimental and Theoretical Physics* reached port,[2] all Soviet publications were simply dumped overboard in the name of defending the free world.[3] The ar-ticle by Ginzburg and Landau — the basis, according to Philip Anderson, of the modern era of superconductivity — thus met an untimely death in the waters of the Hudson River. The paper remained almost totally unknown for a decade. John Bardeen, for example, learned about it only five years later from the microfilm of a mediocre translation that he obtained from David Shoenberg (who had been the liaison between Moscow and Cambridge when Kapitza was prohibited from leaving the USSR).

Physicists on the other side of the Iron Curtain were equally ignorant of what was being done in the West; Western journals could not be found in Soviet laboratories. This was the period when Kapitza was not permitted to work in his own institute. In those areas in which the contributions of the East and the West were known to both camps — in liquid helium research, for example — we have already seen that the tone of the debates did not always correspond to the calm supposed to reign in academic discussions. Thus it was that, during the decisive period for understanding superconductivity, crucial developments occurred on the two sides simultaneously, and barely a word leaked out from one side to the other. Ginzburg and Landau's paper became a real reference point only at the end of the 1950s; in 1950 it was unknown outside the Soviet Union.

Landau had made a first unsuccessful try in 1933 at applying the theory of phase transitions to superconductivity. A similar approach was followed in the article by Ginzburg and Landau, this time in masterly fashion. In fact, Landau's first paper was just wrong, and he had it discreetly removed from his complete works. His mistake lay in choosing the supercurrent density as his order parameter, but he was still convinced that using an order parameter was the best way to approach the problem. After the war he set to work on this in collaboration with his friend V. L. Ginzburg, who worked at the Lebedev Institute in Moscow. It was not far from Kapitza's laboratory, but it had always maintained its independence. Ginzburg was also involved at this time in the Soviet nuclear program, but he was not exactly considered a saint (his second wife had been shipped off to Gorki). In contrast with his colleague Igal Tamm and Tamm's student Andrey Sakharov, Ginzburg was not authorized to visit the "Installation," and this apparently left him time for other pursuits.

In 1950 the two collaborators came up with a new formulation of the physics of the superconducting transition.[4] This time the order parameter was an abstract quantity that the two authors carefully neglected to define with precision, saying simply that it was like a wave function that could represent the superconducting electrons.[5] This abstract formulation did not prove terribly seductive for experimenters when they discovered the paper at the end of the 1950s, and they avoided using it. Making use of both the wave function and the order parameter, Ginzburg and Landau felt free to write down a totally new mathematical expression for the free energy of the superconductor, or, more exactly, for the difference in the free energy between the superconducting state and the normal state.

They really had no basis whatever for thinking that their expression was an adequate representation of the free energy of a superconductor. It was a kind of bet; they said it simply had to be this way. Given an order parameter arbitrarily chosen and a magnetic field, their expression would yield a value for the free energy of the superconductor. The actual values of the two parameters, the order parameter and the magnetic field, had no physical significance. It's a bit like the baker who says, "To make a cake, I'll need some flour, some eggs, and some butter," but doesn't specify the proportions. A neophyte baker would have to make a starting guess at the proportions, then change them little by little and keep trying until what he pulled out of the oven looked like and perhaps even tasted like a cake. It might take some time.

Nevertheless, mathematicians had known about this method for two centuries, and theorists had used it frequently to solve problems in quantum mechanics. In physics it's called the *variational method*; the values of many contributions to the energy are varied until the calculated value of the energy is as small as possible. Happily, there exist mathematical methods that get around the problem of making a huge number of tries; they directly yield the conditions that the parameters must satisfy to minimize the energy.

If the list of ingredients for the cake is not complete at the start, if the butter is left out, for example, the cake is hardly likely to turn out right. When all is said and done, the variational method will turn out a good cake only if the initial ingredients have been carefully chosen. Exactly the same is true in physics. If the original expression for the free energy is not well constructed, if it lacks some fundamental elements, the mathematical methods will yield a result but not one that's likely to correspond to reality. Ginzburg and Landau's stroke of genius was to guess just what had to be put into the expression for the free energy of a superconductor.

Guess — what a strange word to use in describing the formulation of a theory. The reader might well suspect that we've turned to such a word to avoid explaining some complicated process, which is not at all true. We could try, of course, to speculate gratuitously on what transpired between Ginzburg and Landau. Otherwise, we have to admit that here, as in the excitation spectrum of liquid helium, what we have is not the almost mechanical result of a calculation whose form would be accepted right away by almost any physicist. Instead, we have a work of the imagination. Robert Schrieffer, coauthor of the other theory of superconductivity, said as much in a lecture on the Ginzburg-Landau theory complete with all the technical details.[6] The setting was the University of Pennsylvania; it was 1962, and Schrieffer spoke to an audience of specialists: "In 1950 Ginzburg and Landau proposed an extension of the London theory which takes into account the possibility of the superfluid density varying in space. They phrased the theory in terms of an effective wave function. . . . Ginzburg and Landau treated this wave function as an order parameter which is to be determined at each point in space by minimizing the free energy of the system. *The problem is then one of guessing an appropriate form for the free energy.*"[7]

The conditions for minimizing the free energy were expressed in the form of two relations between the order parameter and the magnetic field, which, since 1951, have been called the *Ginzburg-Landau equations*. Fifty years later these equations, often applied in areas (such as nuclear physics) well outside the domain of superconductivity, continue to be exceedingly useful.

The order parameter comes first in the list of ingredients, and it has two unusual properties. On the one hand, it is defined locally, that is, its value can vary from one point to another and thus depends on the coordinates of the point of interest. On the other hand, Ginzburg and Landau found it opportune to make it a complex number, a number with two parts. The energy of the superconducting state is then defined at each point by a mathematical expression involving the order parameter. It's not just the value of the order parameter at the point of interest that matters, however, but also the way the order parameter varies around this point.[8] In fact, the mathematical quantity in the equations is the *gradient* of the order parameter, the mathematical tool (or operator) that expresses the spatial variation of a quantity. As an added twist, Ginzburg and Landau made the order parameter a complex number.[9]

The magnetic field, the second ingredient, also had to be consistent with London's equation, whose predictive power (the Meissner effect, e.g.) made it an unavoidable element of any theoretical approach. Ginzburg and Landau thus introduced the magnetic field into the expression for the free energy in a way that is formally identical to its use in Schrödinger's equation in quantum mechanics, in which it depends on the momentum of the electrons. Given that the momentum of the electrons varies with the gradient of the wave function, Ginzburg and Landau were led to modify the expression for the gradient of the order parameter to take into account the effect of the magnetic field.

Now we have just about all the ingredients of the energy of the superconducting state. The mathematical operation that allows the proportions of these ingredients to change in order to obtain the lowest energy state is expressed by the Ginzburg-Landau equations. The simultaneous solution of these two equations thus provides the values of both the order parameter and the magnetic field in a given situation, and one of the equations even yields the value of the supercurrent.

We've gone on quite a bit about these equations. They represent a true marvel of intuition. What's more, they give the right answers, in remarkable accord with the experimental data; this shows that the chosen ingredients are indeed correct. Of course, it was important that the London equation, with its penetration depth λ_L, could be retrieved from the Ginzburg-Landau equations. In fact, one does find in these equations a length, $\lambda(T)$, that measures the penetration of the magnetic field; λ_L and $\lambda(T)$ are equal at $T = 0$. In addition, one finds a second characteristic length, essential for any order parameter that varies in space, as Brian Pippard had proposed at about the same time; we saw that he cleverly deduced its existence from his experimental data. This second characteristic length, customarily designated by the symbol $\xi(T)$ (pronounced "ksi of T"), represents a distance scale for significant variations in the order parameter.

We can get a concrete idea of what these two quantities mean if we consider a superconducting sample placed in a magnetic field strong enough that a part of the sample becomes normal. The order parameter has a nonzero value in the superconducting region but is equal to zero in the normal region. The situation is just the opposite for the magnetic field, which, except for a short distance near the boundary between the two regions, is zero in the superconducting region.

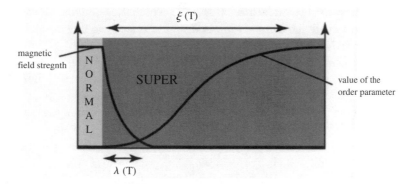

The normal-superconducting interface. The Ginzburg-Landau model of a superconductor allows a description of the surface separating superconducting and normal zones inside the same sample. It predicts that the order parameter that describes the superconducting state increases from zero at the interface to unity over a distance labeled the coherence length $\xi(T)$, while the magnetic field strength decreases to zero over a distance λ inside the superconductor. The ratio $\kappa = \lambda/\xi$ is a parameter characteristic of the material. It determines whether the material is a type I superconductor ($\kappa < 1$, perfect Meissner effect) or type II ($\kappa > 1$, penetration of the magnetic field is possible without destroying the superconductivity).

The magnetic field decreases exponentially to zero with the characteristic length $\lambda(T)$, while the order parameter does the same with the length $\xi(T)$. The transition zone between the two phases constitutes a sort of wall; Landau himself had proposed such a domain boundary to explain what we would call today the intermediate state in type I superconductors. Between two fluids there is an interface with surface tension, an energy per unit surface area characteristic of the interface. By analogy with fluids it was tempting to introduce a surface energy for the domain wall and calculate its value; this was particularly easy to do with the energy expressed in terms of the order parameter. In the situation represented in the diagram, in which $\xi(T)$ is much longer than $\lambda(T)$, the value obtained for the surface energy corresponds well with the value Landau expected from calculations he had done previously.

The Ginzburg-Landau equations are phenomenological equations. They give a general description of the transition region between the superconducting state and the normal state that works for all superconductors provided the two coefficients, the two parameters in the expression for the energy, are chosen appropriately. For a given material, experimenters might measure, for example, the value of the critical field at zero temperature and the London penetration depth. From these data they could deduce the values of the two numerical parameters and so determine $\xi(T)$ and $\lambda(T)$. For most metals and superconducting alloys known in 1950, and, in particular, those used by Pippard, the values obtained for $\xi(T)$ and $\lambda(T)$ were compatible with a transition region between the two states just as we have described earlier, with $\xi(T)$ much larger than $\lambda(T)$. Ginzburg and Landau found it convenient to utilize the ratio of the two quantities, $\lambda(T)/\xi(T)$, which they designated by the symbol κ (*kappa*, now called the *Ginzburg-Landau parameter*), and they concluded that it was always much smaller than unity.

Given the arbitrary nature of the coefficients introduced into the expression for the energy, one might wonder what happens when they take on values different from those corresponding to classic superconductors such as lead, aluminum, and mercury. Of particular interest would be a situation in which $\xi(T)$ is not much *greater* but, rather, much *less* than $\lambda(T)$ — that is, when κ is much bigger than one. According to secondhand information, it would seem that Ginzburg and Landau did consider this. Their result, which appears today in every course on superconductivity, was that the surface energy of the domain wall between a superconducting zone and a normal zone was negative. This means that creating such a domain wall lowers the energy of the superconductor and makes a state that is thermodynamically more stable than the state without the boundary interfaces. To Ginzburg and Landau this situation would have seemed unimaginable, unphysical, and they didn't mention it. Their attitude now seems amazingly cautious, however, especially in comparison with the standard practice among particle physicists, who often don't hesitate to predict the existence of wildly unconventional objects as soon as their equations suggest it. Many examples, antimatter is one, show that they were perfectly right to act this way, but the game is not the same everywhere. In any case Ginzburg and Landau came exceedingly close to the discovery of superconductors with negative surface energies, today called type II superconductors, in contrast with type I superconductors, which have positive surface energies. The irony becomes even richer when we realize that excellent specimens of type II, the best at the time, had been prepared and studied in the USSR, in Kharkov, by Shubnikov.

Perhaps the explanation for this apparent oversight lies simply in the fact that Landau, in writing his article, wanted above all to demonstrate the possibilities of his theory of second-order phase transitions (and, by the same token, fix up his 1933 error). His intention was not to develop a general model of superconductivity but only to prove, in collaboration with Ginzburg, that his theory of second-order phase transitions could be applied to situations in which the order parameter varies from one spot to another; superconductivity merely provided an example.

This vision has turned out to be prophetic. Now, more than fifty years later, the Ginzburg-Landau equations are applied in fields extending far beyond superconductivity. With small changes they are used in a number of situations in which an ordered structure is locally perturbed, for example, in the neighborhood of a fault in the highly deformable crystalline system that constitutes a liquid crystal. Most applications are to be found, however, in the field of superconductivity itself. One of the most famous is precisely the theory of type II superconductors, the product of a student of Landau, Alexei Alexeevich Abrikosov.[10]

Abrikosov was a young researcher at the Institute of Physical Problems in Moscow, the temple of theoretical physics where Lev Landau reigned as high priest and paternalistic despot. Taking seriously the idea that superconductors with negative surface energies at domain walls could exist, Abrikosov thought about how they would behave in an external magnetic field. He knew that creating a wall between normal and superconducting zones lowers the total energy of a superconductor in a magnetic field. Thus, the most energetically favorable state of this system might be reached when the magnetic field penetrates the superconductor, so as to create the largest number of domain walls possible.

We can try to understand this fragmentation with the help of another culinary analogy.[11] Water and oil don't mix, because the energy of the interface between the two fluids is positive. If you try to mix them, they separate, and the surface of the interface between them becomes as flat as possible to minimize the area of contact. Now add some molecules with an affinity for both water and oil. It's easy to imagine that walls made up of a layer of these molecules will form, one side in contact with the water and the other with the oil, representing a state that is thermodynamically more stable. It means that the surface energy at the interface is negative. Such molecules, called lecithins, exist in egg yolks. Lovers of French cuisine know what happens next. With a few other ingredients (salt, mustard, vinegar, etc.) and a good shaking, you'll have mayonnaise; you'll find you need a very sharp eye to distinguish the water from the oil. What's happened is that a system of miniscule spherical droplets has formed, floating in an aqueous solution, each one surrounded by a layer of lecithin. By dividing this way, the surface where the two phases are in contact — the surface of the domain wall with negative energy — has become ten to a hundred thousand times larger than it was before the mayonnaise formed.

When Abrikosov tried to picture for himself a type II superconducting state in the presence of an external magnetic field, he certainly didn't have mayonnaise in mind. In fact, the normal (nonsuperconducting) zones containing the penetrating magnetic field could not exist as closed surfaces like the oil droplets. The laws of electromagnetism (Maxwell's equations) require that the magnetic field lines be continuous and close upon themselves. The magnetic field, however, is generated by external magnets or coils, so the field lines begin there and have to return there; they can't simply end someplace inside the sample. They have to enter one side and leave by the opposite side and traverse the sample from one domain to another. The structure that optimizes the contact between the two zones in this situation is like a bundle of spaghetti, a set of cylinders lined up parallel to each other. The superconducting zones fill the space in between. The axis of each cylinder is perpendicular to the surface where the magnetic field enters, and each cylinder accepts part of the magnetic field.

Abrikosov put his model of the penetration of the magnetic field to the test of the Ginzburg-Landau equations. The solutions he obtained predicted a behavior of the superconductor completely different from any previously known. Instead of the total expulsion of the magnetic field in the superconducting state, the perfect diamagnetism of the Meissner effect, he found a complex situation characterized by a progressive penetration of the magnetic field into the superconductor, leaving the superconducting properties intact.

Step by step, this is what happens when an Abrikosov superconductor feels an increasing external magnetic field. Up to a strength H_{C1} no magnetic flux penetrates. Above this value a finite amount of magnetic flux penetrates all at once, and a new structure is set up, characterized by a regular network of nonsuperconducting filaments separated by large superconducting spaces. Along the axis of the filament the order parameter is zero. As the distance from the axis of the filament grows, the order parameter increases toward the value it has in the undisturbed superconducting zones. The radius of the perturbed region is of the order of $\xi(T)$. The magnetic field lines are squeezed into the little cylinders of normal metal, and the field is greatest along the axis of the normal filaments. The magnetic field penetrates into the superconducting zones over a distance of order $\lambda(T)$, which, for superconductors of type II, is larger than $\xi(T)$. The magnetic field thus decreases with distance from the axis of the filament but less quickly than the order parameter grows. The existence of these tubes of magnetic field arises from the circulation of supercurrents around the axis of the filaments; supercurrents flow just like water running in little whirlpools down a drain and, like them, are called *vortices*.

Each vortex is a well-defined entity, and the magnetic flux that it carries is not arbitrary. Its magnitude is fixed by the same quantum mechanical principles that determine the orbits of electrons around atoms. They lead to the result that each filament contains one quantum of magnetic flux, equal to h/2e, or $2 \times 10^{-15} \mathrm{Tm}^2$.

It helps to think about this in the following way. Far from the axis of the vortex, the order parameter has its superconducting value, and the magnetic field has really dropped off to zero. We would think that nothing should distinguish the superconducting state in this region from what it would be in the absence of a vortex. But this is not quite true. Even if the magnetic field is zero far from the axis, the properties of space are slightly changed by the presence of a nonzero field at the center of the vortex,[12] in such a way that the order parameter is not truly constant from one point to the next. Remember that the order parameter is a *complex number*, a number with two parts, which here we can consider as an amplitude (a magnitude) and a phase (an angle). The amplitude, a measure of the superconducting electron density, is not affected by this modification of the space, but the phase is sensitive to it and so varies from point to point. After a complete turn (360°, or 2π radians) around a vortex, the point in space is exactly the same as the starting point, so the order parameter must remain the same. The same is true for an integer number k complete turns. The phase variation around a complete turn depends on the magnetic flux contained within the encircled area in a way specified by the Ginzburg-Landau equations. The result is that the flux surrounded in one complete turn has the value $\Phi_o = h/2e = 2 \times 10^{-15} \mathrm{Tm}^2$. After k turns, the flux surrounded is $k\Phi_o$.

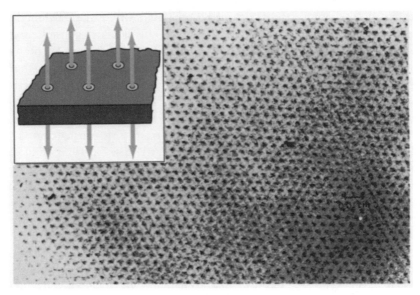

The Abrikosov vortex network in a type II superconductor. The description of the mixed state of type II superconductors proposed by Abrikosov in 1952 represents one of the most spectacular successes of the Ginzburg-Landau model. Abrikosov suggested that the magnetic field would penetrate the sample inside flux tubes (vortices) regularly spaced in a triangular array. This structure was experimentally confirmed twelve years later by covering the surface of the superconductor with a fine ferromagnetic powder that was concentrated on the tubes as shown. The dynamics of this network of vortices, and its anchoring in faults in the crystal structure, play a primordial role in the electrodynamic properties of these superconductors and affect all their applications. The insert is an enlarged view of the lattice of vortices; the arrows indicate possible directions of the magnetic field (never both at the same time).

This property, known as *flux quantization*, had been predicted by London in 1938, not for the magnetic flux carried by a vortex but for the flux in a superconducting ring carrying superconducting currents. No experimental measurement had been able to verify this prediction, but this was not surprising, given the tiny value of the flux quantum. Had such an experiment been successful, it would not have verified London's prediction exactly, since the prediction was wrong by a factor of two. London's reasoning was based on the variation of the phase of the wave function of an electron around a ring; this yields a value for the elementary quantum of flux h/e. In fact, it is not the wave function of an isolated electron that should be calculated but, rather, the wave function of a pair of electrons, a Cooper pair, as we shall discover in chapter 16. This leads to the replacement of e by 2e in the expression for the elementary quantum of flux.

Abrikosov found that each vortex carries one and only one quantum of flux, an important result. As the magnetic field starts to penetrate the superconductor, just above H_{c1}, the vortices form a rather loose lattice, but, as the field increases, they get closer and closer together, and the lattice becomes tight with a triangular mesh. The mean distance between vortices approaches $\xi(T)$. The paradoxical result predicted by Abrikosov is that the sample is then entirely penetrated by a magnetic field of considerable strength (several tens of teslas for certain superconductors), all the while remaining a superconductor. The superconducting state is finally destroyed when the field reaches a value traditionally labeled H_{c2}. At least, that's what Abrikosov predicted.

Abrikosov remembers that he conceived of this surprising structure and made his calculations shortly after the publication of the article by Ginzburg and Landau, around 1952. When he submitted his manuscript to Lev Landau, Landau supposedly rejected it and forbade his student to publish it. According to Abrikosov, he had to follow his professor's advice; he put his manuscript in a drawer and kept his mouth shut—until an article by Richard Feynman appeared in 1954 predicting the existence of vortices in a container of rotating superfluid helium.[13] The similarity to his rejected paper was such that he dared to confront Landau once more, and this time Landau enthusiastically encouraged his disciple to publish it. So goes Abrikosov's version of the events. In the meantime Stalin had died, and it became possible once more to mention the names of his victims, to cite their work, or at least to read them (it was better to cite a prisoner than a foreigner). Thus, Abrikosov was able to remind people of the magnetization curves of certain superconducting alloys obtained by Shubnikov at Kharkov in 1938, curves that showed the penetration of the magnetic field in a superconductor in exact accord with his vortex model.

If the circulation of Ginzburg's and Landau's paper outside the Soviet empire was limited to the dark green waters of the port of New York, such was not the case with London's book. The title, *Superfluids*, summarizes with its plural the ambition that animated it.[14] He wanted to push as far as possible the analogy between superconductivity (vol. 1, which appeared in 1950) and the superfluidity of helium, the subject of the second volume, which appeared in 1954 just before he died. To construct this analogy, London systematically reviewed all his work from before the war and everything that had just been established in the discussions about superfluid helium. This kind of work often has a self-serving aspect, out of sync with the times. Fritz London is not entirely above reproach in this regard, but the immediate response to the book clearly showed that he had not been well understood before then. The efforts of some very illustrious theoreticians reveal the persistence of earlier obsolete ideas in their attempts to explain superconductivity.

As an example, in 1948 Heisenberg, accepting the risk that he might seem old-fashioned, had looked for a way to describe superconductivity in terms of electron crystals, and so he studied the electrostatic forces between the electrons.[15] For Heisenberg it was zero resistivity that characterized superconductivity, whence the suggestion of geometric order. A crystal of electrons sliding smoothly across the ionic lattice (and parallel to one of its principal axes) would, in fact, explain the lack of resistivity, but there was no way that it could describe the Meissner effect. It was as if

London's work had never happened.[16] London replied sharply that it was perfectly acceptable to take the interactions between the electrons into account in treating superconductivity, but he could not share Heisenberg's point of view. For Heisenberg he said, "The essential difference . . . is the hypothesis that it is perfect conductivity rather than diamagnetism that is the principal characteristic of the phenomenon."[17]

For London superconducting electrons and superfluid helium are both quantum liquids:

These strange fluids are a challenge to the theoretical physicist. The mere fact that they are non-solid states, probably right down to 0°abs, precludes an explanation of their structure on the basis of classical mechanics. According to the classical theory, at a sufficiently low temperature, any system composed of identical particles confined to a given volume should come to rest (vanishing kinetic energy) in the configuration of minimum potential energy. This obviously would be some structure such as a crystal lattice, which is ordered in space; in short, a solid and not a fluid state.

This explains why:

the existence of the superfluids and in particular the strange transport mechanisms they exhibit are indirect indications that they represent macroscopic systems for which the classical theory is incompetent and that presumably quantum mechanics is relevant to their constitution *as a whole* [italics of F. London here and below]. . . . We mean that they are, although of microscopic size, nevertheless withdrawn from the disorder of thermal agitation in essentially the same manner as is the electronic motion within atoms and molecules, in short, that they are *quantum mechanisms of macroscopic scale*.[18]

London had stated each of these propositions previously. Placed end to end like this for the first time, however, they provided a fresh view of the whole subject, a view that only a small minority of insightful theorists had perceived previously. London insisted on the fact that his theory, because it led to an understanding of the perfect diamagnetism of a superconductor, presumed that the superconducting electrons were in a macroscopically ordered state — that is, ordered over large distances compared to the scale of atoms.[19]

This "order" was not like that of a crystal in ordinary space; London said it was ordered in momentum space, not in position space.[20] The wave functions of the superconducting electrons are, according to his expression, rigid. What he meant was that they are the same everywhere in the superconductor and are independent of the applied magnetic field. The expression for the penetration depth given by his theory is indeed independent of the strength of the applied magnetic field, as experiments had already confirmed.

Now position and momentum are related by the famous Heisenberg inequality, popularly known as the uncertainty principle. It tells us that the position and the momentum of a particle cannot both be known with certainty at the same time. The product of the uncertainty in each quantity must be larger than Planck's constant h. London knew from electromagnetic theory that the momentum of the superconducting electrons was roughly the same everywhere in the superconductor. So, he reasoned, the uncertainty in the position must be very large, meaning that the wave function of each electron must extend over a large distance: the wave functions of all the electrons must overlap one another to form a macroscopic quantum state.

London recognized well enough that he did not yet know how this long range momentum space ordering was set up. What he did know was that it didn't require that the superconducting state be a fixed geometric configuration (i.e., a crystal) with an energy near the minimum of the potential energy. Precisely because of Heisenberg's inequality, he said, the kinetic energy must also be involved. That's why "the electrodynamics of superconductivity furnish a *definitive reduction* of the problem posed for a future molecular theory." (For London the expression *molecular theory*, which is no longer used, meant a quantum theory that explains molecular structures, be they microscopic or macroscopic.) "No longer does it seem necessary to explain why the electrons in superconductors lose all vestige of resistance. The task will be to show that they prefer to 'solidify' with respect to their momenta rather than with respect to their coordinates."[21] Heisenberg was not cited, but the allusion is clear: London did not want to miss the opportunity to contrast the former head of the Nazi nuclear program to von Laue, who had had the courage to cite London during the war.[22]

With *Superfluids* London thus reaffirmed the point that it is the hydrodynamic approach to quantum liquids that provides the royal path to the understanding of superconductivity. He didn't suggest it was the only way, because he was well aware that his starting point was purely phenomenological. He emphasized that his approach was conceptually more interesting. The success that he felt he had achieved with superfluid helium gave him confidence in his approach, as did the results of the experiments on superconductivity. After the work of his brother and of Shoenberg, the penetration depth was measurable; it was independent of the applied magnetic field, and it had the predicted order of magnitude.

London, even if he could properly feel that he had not been understood, was thus not a lone warrior. Was the hydrodynamic approach going to be able to supply a microscopic explanation of superconductivity? One theorist was particularly well positioned to answer that: Fritz London himself. Unfortunately, London died of a heart attack in 1954 at the age of fifty-four. His passing, in the same year that the second volume of *Superfluids* (devoted to helium) appeared, deprived the theater of superconductivity of a crucial actor.

CHAPTER 15

Now, How to Grab the Tiger by the Tail?

Ginzburg and Landau were ignored; London had passed away. It was three theorists on the other side of the world, at the University of Sydney in Australia, who now led the attack on the problem of superconductivity from the hydrodynamic perspective. M. R. Schafroth, a spirited former student of Wolfgang Pauli, was the head of the group. A Swiss citizen, he had been convinced to emigrate in 1954 by Harry Messel, a professor at the University of Sydney who wanted to establish a real school of physics there. Having convinced several millionaires dazzled by the promise of big science to provide funding, Messel transformed himself into a recruiting sergeant and tried to attract the brightest young physicists.[1] This was the era when the afterglow of the atomic bomb still provided a dubious halo of prestige to physics. Schafroth had just spent a year with Herbert Fröhlich in Liverpool and now worked with John Blatt and Stuart Butler. Blatt was a former student of Weisskopf (whom we met in Kharkov and will soon meet again); Butler had worked with Rudolf Peierls. The trio were sheltered in a laboratory devoted to nuclear physics and computation. Although this was Australia, they weren't quite so isolated as it might appear; Schafroth was in contact with Pauli and Fröhlich in Europe. Their productivity was extraordinary: five of their articles, one after the other, appeared in just one issue of the *Physical Review* (October 15, 1955). All dealt with the properties of a gas of bosons, particles obeying Bose-Einstein statistics.

Schafroth was the sole author of the two papers directly concerned with superconductivity, no doubt because, in a previous publication, he had suggested describing the superconducting state as a perfect gas of charged bosons.[2] His reasoning was simple:

So far, no molecular theory of superconductivity has been found. The most successful attempts in this direction have been made by Fröhlich and Bardeen on the assumption that the occurrence of this phenomenon is due to the interaction of the conduction electrons with lattice vibrations. . . . However, it has not so far been possible to show that a strong enough lattice-electron interaction can account for the characteristic equilibrium phenomena of superconductivity, namely the phase transition and the Meissner-Ochsenfeld effect.

In fact, even though Fröhlich and Bardeen had recognized that an electron-phonon interaction was necessary to explain superconductivity, their theories ran into insurmountable mathematical difficulties. Schafroth had been the first to realize that it was impossible to explain the Meissner effect as long as the electron-phonon interaction was treated as a small perturbation on the system.[3] This is not the case, he said, with a boson gas, which can serve as "a guide in the search for the real phenomenon." Schafroth emphasized that this is because "Bose-Einstein condensation is a singular and unique phenomenon that is responsible for the other spectacular effect in low temperature physics: superfluidity. It thus seems reasonable to suppose that superconductivity in metals is due to the appearance of charged bosons in the metal." After citing all the leading lights of the hydrodynamic approach—London the brightest among them—Schafroth then raised the crucial issue: What can behave like a boson in a superconductor? Particles that obey Bose-Einstein statistics must have integer spin, and the electron has spin $\frac{1}{2}$. To have integer spin it is sufficient, however, simply to consider two electrons as a single entity. Because the two have the same charge, we're now dealing with doubly charged bosons.

This suggestion was a direct connection to London's work on superfluidity. The nature of the coupling between the electrons was not defined, but Schafroth was thinking about some kind of resonating state of two electrons (the language was borrowed from chemistry). He was thus making a molecular or microscopic proposal for describing superconductivity, but he didn't go into detail about how the bosons were created. In fact, there's an overriding difficulty: electrons all have the same charge, and so they repel one another. The mechanism responsible for creating bosons would thus have to overcome this repulsion. Imagining such an attractive force proved a formidable hurdle, and Schafroth went no further. Incomplete as it was, Schafroth's contribution was the first try at explaining superconductivity through electron pairing. In addition, he succeeded in describing the essential characteristics of a superconductor: a well-defined temperature where the phenomenon appears, a Meissner effect, and a penetration depth for the magnetic field. It did not, however, predict the discontinuity in the specific heat at the critical temperature or the proper variation in the magnetization as a function of the applied field.[4] The gas of charged bosons thus was clearly not *the* microscopic solution of superconductivity, but Schafroth remained convinced that he was on the right path; he expected that "a more refined elaboration of the same picture would lead to appreciable improvement."

The Australians wanted to develop a complete theory starting from what they called a chemical quasi-equilibrium, but they ran into mathematical problems. They did succeed, however, in explaining qualitatively what they were after. Their bosonic electron pairs were small objects separated from one another by distances much greater than their size, so they could move around without bumping into one another. As Schrieffer later remarked, repeating what he had no doubt said much earlier, such a system would have yielded a continuous spectrum above the Fermi energy and thus no energy gap.[5]

The Australian group clearly had the impression that the harbor was in sight, and so they worked at full steam throughout the 1955–1956 academic year. Until then superconductivity theory had been pursued by brilliant individuals working more or less alone. The Australians worked as a team; they were visibly enthusiastic and no doubt a bit arrogant; the great masters encouraged them. In addition, as often happens when a field matures, superconductivity was becoming fashionable. An international conference on theoretical physics was supposed to take place in Seattle in September 1956, and the organizers had asked Richard Feynman, fresh from his distinguished work on helium, to present a lecture on superconductivity. Blatt, one of the Australian trio, was in the audience.

Feynman at this time did not yet have his Nobel Prize, but physicists universally considered him exceptionally smart. One of his most notable achievements was a remarkably elegant and efficient method for solving problems with a large number of interacting particles, the "many body problem." His "Feynman diagrams" are still in constant use in diverse fields, especially nuclear and particle physics.[6]

Feynman was an astute choice for a critical review of the theoretical situation in superconductivity. He had made an important contribution to understanding the superfluidity of helium, showing that energy dissipation in superfluid helium takes place primarily via quantized vortices.[7] For superconductivity, in which no theoretical attempt at a solution seemed to be succeeding, a speaker whose interests were a bit removed from superconductivity itself could provide a useful external perspective. The title of Feynman's talk at the Seattle conference was all-encompassing: "Superfluidity and Superconductivity."

Feynman first deplored Landau's absence; as usual, the Soviet authorities had no doubt refused him a visa the day before his scheduled departure. Feynman then described the glory days of Leiden. His chronology was, shall we say, eccentric, but his confusion was unlikely to shock his listeners, who were not yet accustomed to the introduction, since become a ritual, of the sort: "In 1908 Kamerlingh Onnes liquefied helium. . . ." In any case the audience was not there to judge whether Feynman was a good historian.

The core of his presentation started with a summary of the results he had obtained over the past two years on helium vortices; he then went on to superconductivity. He mentioned, of course, the model of Fröhlich and Bardeen, which proposed an electron-phonon interaction, as well as the improved version of Bardeen and Pines that took into consideration the Coulomb repulsion.[8] For Feynman, if superconductivity was not well understood, it was because the problem was not well formulated and no one had yet been clever enough to do the mathematics. Having mentioned that he also had spent some sleepless nights thinking about these issues, he suggested that one pick out some aspect of superconductivity, anyone at all but just one, and try to explain it. Alongside this obvious pragmatism, however, Feynman also claimed the high road; any explanation had to be deduced from first principles, the greatest ambition a theorist can have. In this case first principles meant Schrödinger's equation, with all the ingredients of the problem properly included. Then some property of superconductivity should be described:

> I would like to maintain a philosophy about this problem which is a little different from usual: It does not make any difference what we explain, as long as we explain some property correctly from first principles. If we start honestly from first principles and make a deduction that such and such a property exists — some property that is different for superconductors than for normal conductors, of course, — then undoubtedly we have our hand on the tail of the tiger because we have got the mechanism of at least one of these properties.[9]

Feynman's concept was just the opposite of London's phenomenology, for whom diamagnetism was the crucial property that had to be explained. Feynman was perfectly happy to recognize that diamagnetism was *the* characteristic of superconductors. Nevertheless, in preparing his talk, he had started from the work of Fröhlich and Bardeen, which was already several years old, and he had labored several months on superconductivity without getting any further along than the others. What element was missing from the puzzle of superconductivity? His conclusion could not be challenged — the method had to be changed: "I am now brought to the same position as Casimir, who first told me about this problem. He said, 'There is only one way to go about working this out. It is simply to *guess* the quality of the answer.' . . . The only reason that we cannot do this problem of superconductivity is that we haven't got enough imagination."[10]

These words were barely out of his mouth when the audience became transfixed by an event vividly described by P. W. Anderson:

> The most dramatic moment I remember having to do with superconductivity came at the end of Feynman's talk at the Seattle theoretical physics conference in September 1956, when he announced that, although he had solved the problem of superfluidity, he had spent many months computing on the problem of superconductivity and had failed utterly. At this point John Blatt leaped up on the stage and announced: "We *have* that idea, and we *have* solved the problem." The idea was pairs, indeed, but the Australian group was very far from a formal solution of the problem.[11]

By the end of the Seattle conference Bardeen's student Robert Schrieffer felt quite relieved because nothing decisive had been said concerning his thesis topic.[12] His subject was the theory of superconductivity, and he had been working on it for a year. It was not so much the Australians whom he was afraid of as Richard Feynman himself; he was not concerned at all about John Blatt, who had just caused all the commotion. Blatt had stopped in Illinois, where Schrieffer had been working with Bardeen, and they had had numerous discussions, as Blatt tried to convince Bardeen that what the Australians called the quasi-chemical method was the right one. Bardeen had not changed his opinion; for him their hypothesis was physically unjustified. On the other hand, Schrieffer had heard it said that Feynman, on a stop in Tokyo, had filled many blackboards with an enormous number of diagrams. As he would explain in Seattle, he was looking for a singularity in the specific heat that could be used as a marker of the presence of superconductivity. Just as Bardeen had suspected, it finally turned out that the efforts of the Australians collapsed; ultimately, the solution did not come from them or from Feynman either.

Six years had gone by since the discovery of the isotope effect and the first models with an electron-phonon interaction. In contrast with the theory of helium, theoretical progress on superconductivity was stalled, and this was not for lack of hard work. What was the missing ingredient that would allow them to grab the tiger by the tail? At the time of the Seattle conference Schrieffer and Bardeen already seemed to know. The end of John Bardeen's long hunt was already in sight.

CHAPTER 16

John Bardeen's Relentless Pursuit

Three years after the discovery of the transistor effect, for which he and his two collaborators won the Nobel Prize,[1] John Bardeen, a homegrown product of American physics, published the first of his articles on the isotope effect. For him this was not so much a change in focus as a return to a previous interest after a long hiatus. Like Landau, Bardeen had been a child prodigy in mathematics. He was born in 1908 to an academic family. He left primary school for high school at the age of nine but continued to play after school with friends of his own age. In spite of his obvious talents in physics and mathematics, Bardeen majored in electrical engineering at the University of Wisconsin and received his bachelor's degree in 1927. Gulf Oil hired him for its research laboratories, where he developed an electromagnetic method for detecting oil deposits so new that Gulf decided not to patent it lest its rivals find out. The method was revealed only thirty years after the fact.[2]

Not seduced by his large salary, Bardeen decided after a few years to return to school to pursue a Ph.D. degree in theoretical physics. The news that Einstein would be going to Princeton settled his choice of university. When he got there, however, he learned that Einstein had in fact gone to the Institute for Advanced Study, next door to, but not a part of, Princeton University, and had no interest in having graduate students. Finally, Eugene Wigner, a future Nobel laureate, became his thesis advisor.

With his previous student Frederic Seitz, Wigner had published a decisive article on the work function of sodium, that is, the amount of energy it takes to extract an electron from its surface. He then set John Bardeen to the task of generalizing their result and finding a way to calculate the work function of sodium at zero temperature starting from first principles. The subject itself was not particularly exciting, but it required considerable care and sophistication in the calculations.

Bardeen's work had to be done in two stages; having a look here at the basic ideas is relevant to what follows. In the first phase it's only the properties of the interior of the solid that are important. Bardeen had to calculate what happens to the lowest state of the system, the ground state of the metal, when an electron is taken out of the metal and moved to a hypothetical spot where the electric potential is the same as the average electric potential felt by the electrons inside the metal. This was roughly what Wigner and Seitz had done for an infinite perfect crystal. The second stage was much less obvious. Bardeen had to calculate the difference between the real potential infinitely far from the metal and the average potential in the interior. This difference affects the distribution of charges at the surface, so he had to find a way to take into account what was happening with the electrons near the surface. Bardeen assumed that the actual distribution of positive charges carried by the metal ions in this region could be replaced by a uniform distribution with a sharp boundary. Nowadays this model is called "jellium," as in jelly. Bardeen was not the first to simplify things in this way, but his thesis was the first to yield realistic results.

Electrical neutrality is maintained in the interior of a metal with an exact balance of the positive charge of the ionic lattice and the negative charge of the electrons. Near the surface, however, this balance is not perfect; the positive charge ends abruptly at a sharp surface, while the negative charge drops off more smoothly. The result is a net excess of positive charge just inside the surface and a net negative charge just outside. The separation of net positive and negative charge constitutes what is called an electric dipole and causes a potential difference across the surface that Bardeen had to calculate. He showed that the dipole potential had to include an extra term beyond what had been considered previously.

Connoisseurs might want to continue a bit further. The calculational method that Bardeen used was, in fact, the Hartree-Fock method of self-consistent fields. This technique was becoming quite fashionable for calculating electron densities in atoms. The basic idea proposed by Hartree was to consider the electrons as moving independently of one another in a potential arising from the sum of all their Coulomb fields. The extra effect of one individual electron on another nearby electron was neglected. In 1930 John Slater in the United States and Vladimir Fock in the Soviet Union showed that one could include the effects of the Pauli principle (that no two electrons of the same spin can be in the same quantum state) by adding an extra term, called the *exchange potential*, to the Hartree potential. The exchange potential effectively caused some attraction between the electrons and lowered the interaction energies. Bardeen went beyond this Hartree-Fock method. To achieve better agreement with the wave functions and thus the electron densities near the surface, he had to add another term to the potential, called the *correlation potential*. This took account of the fact that the approximation of purely independent electrons was not good there; their strong mutual repulsion caused correlations in

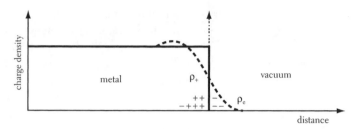

The distribution of electric charge in a metal. The diagram is a simplified version of the distribution of positive charge (ρ_+, solid line) and negative charge (ρ_-, dashed line) in a metal. The sharp cutoff of positive charge at the edge of the metal is characteristic of the *jellium* model of a metal. The total charge of the metal is zero (the metal is electrically neutral), but the electrons spread out slightly past the surface, leaving a small excess of positive charge just inside. This separation of positive and negative charge at the surface consitutes an electric dipole, which Bardeen was the first to calculate realistically.

their positions. Outside the metal all these potentials converged to the same value.[3]

In the midst of calculations for his thesis and publications on diverse topics, Bardeen made an important calculation from first principles of electron-phonon scattering in a metal.[4] The resulting paper proved to be the beginning of Bardeen's relentless pursuit of the theory of superconductivity.

Two opposite approaches to the problem of electron-phonon scattering had been considered previously. The one supposed that the ionic lattice was rigid and the other, that it was deformable; neither yielded satisfactory results. Bardeen reconciled the two approaches by calculating the deformation of the spatial distribution of the valence electrons, the ones that stay attached to the atoms as the lattice vibrates, and he got much better results.

It was only a question of time before Bardeen's interest in the electrons of a metal led him to superconductivity. He had long discussions with John Slater about the work of the London brothers. At about the same time, Shoenberg's book appeared, the first on superconductivity in English.[5] After the publication of a first article in 1940, however, the war took Bardeen away from superconductivity. His geological experience led to a position at the Naval Ordnance Laboratory in Washington from 1941 to 1945, where he studied the problems posed by magnetic mines.

Bardeen wrote an article in 1950 advancing the idea of interactions between electrons and phonons;[6] it was followed by several others. David Pines stands out among his collaborators; the two worked side by side for thirty-two years.[7] His impression of Bardeen's working habits thus has special meaning:

Focus first on the experimental results, by careful reading of the literature and personal contact with members of leading experimental groups.

Develop a phenomenological description that ties the key experimental facts together.

Avoid bringing along prior theoretical baggage, and do not insist that a phenomenological description map onto a particular model. Explore alternative physical pictures and mathematical descriptions without becoming wedded to a specific theoretical approach.

Use thermodynamic and macroscopic arguments before proceeding to microscopic arguments.

Focus on physical understanding, not mathematical elegance. Use the simplest possible mathematical descriptions.

Keep up with new developments and techniques in theory, for one of these could prove useful for the problem at hand.

Don't give up! Stay with the problem until it's solved.[8]

In fact, Bardeen followed all these precepts in his decisive work on superconductivity, the lack of mathematical elegance included.

It quickly became evident both to Fröhlich and to Bardeen that the description of the electron-phonon interaction that they had proposed was not satisfactory. It provided a good explanation of the isotope effect, but the energy difference between the normal and the superconducting states was much larger than experimental estimates. So Bardeen launched an attack on the problem in four seemingly different directions. With hindsight we can note only that his attack converged on the microscopic theory of superconductivity, even though the results obtained in each of these thrusts were not directly used in the eventual theory.

An invitation to write the article on the theory of superconductivity for the *Handbuch der Physik* of 1955 allowed Bardeen to develop a phenomenological description of the main experimental facts. This was the first line of attack. Previously, Bardeen had encouraged Pines to study strong interactions between electrons and phonons, in the expectation that they might turn out to be useful. Pines thus began to study polar crystals, in which the electrons are strongly coupled to high-frequency phonons, forming a combination that acts like a particle and is labeled a *polaron*.

Attack number two: working with Francis Low and T. D. Lee, two theorists who worked primarily in particle physics, Pines developed a variation on a technique that had already been used in the study of mesons. He applied it to the calculation of the properties of solids, in which the electron-phonon coupling was intermediate between strong and weak.[9]

The third direction: the following year, Bardeen wanted to figure out what effect the Coulomb repulsion between electrons had on the electron-phonon interaction. To do this, he had to learn some new mathematical techniques. He then confirmed his results,[10] from 1937, on the phonons in sodium and generalized Fröhlich's results by including the shielding effect of the Coulomb repulsion. This model yielded a curious prediction: two electrons in a solid were not condemned to always repel each other. It seemed possible that under certain conditions they might attract each other.

Finally, with Pines's postdoctoral fellowship coming to an end, Bardeen considered an attack from a fourth direction: a return to the famous energy gap that had been missing in all his approaches to superconductivity based on electrons interacting via the intermediary of phonons. Bardeen felt that he would need new methods to resolve this problem. He started to look for a bright young theorist who would master the field theory techniques in which Feynman had distinguished himself. Chen Nin Yang, from the Institute for Advanced Study in Princeton, recommended Leon Cooper, who arrived in Urbana to work with Bardeen at the University of Illinois in September 1955.

Two years earlier Bardeen had accepted as a thesis student Robert Schrieffer, who had just completed a bachelor's degree in physics at MIT. Schrieffer spent his first days as a theorist working on semiconductors. Bardeen suggested that he pass his second year in the semiconductor laboratory that Bardeen directed in parallel with his theoretical group; Bardeen thought this might be a good introduction to the ways of experimental physics. The result was not long in coming: Schrieffer singed his eyebrows trying to solder in an atmosphere of hydrogen. That was that. Bardeen recommended a return to theory and gave him a list of ten possible thesis topics, the last of which was superconductivity. Francis Low, who had worked with Lee and Pines, encouraged him to start there.

Thus it happened that, as the school year 1955–1956 began, the trio that would later be widely known by the acronym BCS took shape. Cooper started by giving a series of lectures to the other two on *Green functions*, mathematical expressions used in quantum field theory. He emphasized the use of Feynman diagrams, which provided a very powerful schematic representation of the interactions between particles. For his part Bardeen discussed his idea of the work they had to do. Like London earlier, he thought that superconductivity resulted from the long-range order of a quantity that remained to be determined but which in any case had to correspond to an ordering of the electrons in momentum space. All the electrons of the superconductor had to form a system in a single quantum state over a large area. Bardeen also stressed that a real energy gap, with no states at all at low energies or at least a strong decrease in the density of low-energy excited states, had to be an essential characteristic of a superconductor. This followed from experimental results for the electronic contribution to the specific heat capacity as well as from measurements of surface impedance and optical reflectivity. The gap had to be several orders of magnitude smaller than the energy of each electron. It was hopeless to look for it by measuring the difference between two energies — after all, you wouldn't set about measuring the weight of a feather by using a bathroom scale to determine the difference in your weight with and without the feather. The essential elements of the physical mechanism had to be isolated.

A glimmer of light began to shine in the spring of 1956. Cooper studied the interaction between two electrons at the top of the Fermi sea; the electrons in the sea were moving in random directions, but the sea as a whole was motionless. What he found, a real surprise, was that, if the interaction between the two electrons was attractive, no matter how weak, they formed a state in which they were bound to each other. Thus, a finite amount of energy had to be supplied to break up the pair and make the electrons once more independent of one another. This was the first indication of how an energy gap might come about.[11] In other words, Cooper found that pairs would form as soon as there was some kind of attractive interaction between the electrons. Bardeen was immediately convinced that this was the key to the puzzle. All three then tried to generalize the problem by treating all the pairs at once, instead of just a single pair. Starting from simple arguments based on energies and the Heisenberg inequalities, Bardeen showed that the size of a pair had to be very large compared to the average distance between pairs. In fact, this size turned out to be the analog of Pippard's coherence range.

Bardeen was convinced that he now had to add the electron-phonon interaction to Landau's model for the excitations of an electron gas in its ground state. Landau had described this ensemble of excitations as a liquid whose particles did not interact with one another and obeyed Fermi-Dirac statistics; obviously, this model neglected the crucial interaction responsible for superconductivity.

The entire autumn of 1956 was devoted to investigations of a mathematical and physical framework that would make such calculations possible. During this period Schrieffer had the opportunity to observe his mentor closely. Bardeen was extremely introverted; during the course of a conversation he might be quiet for ten minutes or more. If Schrieffer decided that this was a signal that the conversation had ended and headed for the door, Bardeen would emit a few words that he couldn't quite catch. So, Schrieffer developed a strategy—he would say something slightly wrong. That got the conversation moving again.

Such meetings went on without any significant progress, but Bardeen, in contrast to Schrieffer, was not worried. Knowing that Cooper had found the key, he was not afraid that the Australians, who were hooked on bosons, were on a fast track toward a solution. On the other hand, Feynman, who would otherwise have been a formidable competitor, had revealed in Seattle that he was indeed working on the problem but had not found a way to make rapid progress. In fact, there were pairs, and there were pairs. The pairs of Blatt and Schafroth, as we shall see, were not comparable to Cooper's pairs, and Feynman, well aware of the Australian pairs, was not apprised of what was going on in Urbana.

What Schafroth thought would work for superconductivity was a brute force transfer of the Bose-Einstein condensation methods that had worked so well for superfluidity. The problem was that electrons obey Fermi-Dirac statistics and not Bose-Einstein statistics, so they cannot all condense into a single state. He was driven to fabricate bound states of two electrons with opposite spins. These new objects, with spin zero, obeyed Bose-Einstein statistics and thus could condense. Schafroth had proposed putting pairs of electrons together to form what the Australian group called localized molecules. It then became possible to imagine a condensed phase with a perfect Meissner effect. The difficulties they encountered, which were always charged to their mathematics account, would today suggest a dead end. That was far from the case at the time, an era when few theorists dared approach the problem.

Now what exactly is the difference between Cooper's pairs and Schafroth's pairs, such that the one brought ultimate success and the other led no-where? To understand the problem, you have to remember that electrons are quantum objects that sometimes behave like particles, tiny grains of negative charge whose positions are well known when they hit the TV screen, and sometimes like waves, with a particular wavelength and a di-rection of propagation. Then it's the electron's speed that is known, and this is tied in a simple way to its wavelength. Quantum mechanics tells us that when one of these quantities, the position or the speed, is precisely defined, the other is practically unknown. Thus, the electron wave in a metal is what we have called a *Bloch wave*, a wave that spreads out over the entire metal and takes up a lot of space for a single electron. In exchange for this total "delocalization," this lack of knowledge of the electron's position, the mo-mentum (the product of its speed and mass) can be defined almost exactly.

To make a Cooper pair, consider two Bloch electrons as associated with each other; each has a well-defined speed but poorly defined position in the metal. The only requirement is that they attract each other — and this is easier if they aren't very close to each other, because separation re-duces the repulsive effect of their common charge. The momenta of the two participants in the pair must be aligned almost exactly opposite each other, so that the speed of the pair is as close to zero as possible. This pair is a new quantum object with a precisely defined momentum and, there-fore, poorly defined position. The pair is delocalized in a domain whose di-ameter is of the order of 10^{-6} meters, that is, several thousand times the distance between two atoms in a metal.

Schafroth's pair, on the other hand, is a kind of molecule made of two electrons bound to each other but close together in space. Physics jargon has it that Schafroth's pairs arise from condensation in position space, whereas Cooper's pairs arise from condensation in momentum space. Con-sidering the large number of electrons that condense as pairs and the di-mensions of a Cooper pair, it is clear that Cooper pairs must overlap, that each point of space can simultaneously be assigned to a very large number of pairs. It is the result of this overlap, this conglomeration of pairs tangled up in each other, that constitutes the superconducting state that Bardeen and his group tried to construct during the winter of 1956. The entire col-lection of electrons in the metal can then act together to form the macro-scopic quantum state that London had foreseen. Schafroth's pairs, on the other hand, are tiny and do not overlap; it is very difficult to construct a macroscopic state out of them. By way of comparison, it's clear that it is much simpler to build a bridge with girders several tens of meters long of reinforced concrete than with bricks alone.

Cooper sent an article, a letter to the *Physical Review* which was published on November 15, 1956; it had been received on September 21, 1956, just a few weeks after the conference in Seattle.[12] Schrieffer had already been talking about its contents since the spring,[13] in his homage to Bardeen, and he also spoke about "those Feynman diagrams." In his publication Cooper emphasized the range of his work and its conceptual value: "Because of the similarity of the superconducting transition in a wide variety of complicated and differing metals, it is plausible to assume that the details of metal structure do not affect the qualitative features of the superconducting state. Thus, we neglect band and crystal structure and replace the periodic ion potential by a box of volume V. The electrons in this box are free except for further interactions between them which may arise due to Coulomb repulsions or to the lattice vibrations." This did not make Schrieffer any more optimistic. Cooper and Bardeen had each made significant contributions over the past year, whereas he himself had still done nothing that might have allowed him even to hope to finish his thesis. In desperation he turned to a problem in ferromagnetism. On the eve of Bardeen's departure for Stockholm to pick up his first Nobel Prize, he had to encourage his student, telling him that he felt they were close to the crucial step.

In January 1957 Schrieffer took part in a small theoretical workshop on many-body problems. There it occurred to him that "because of the strong overlap of pairs perhaps a statistical approximation analogous to a type of mean field would be appropriate to the problem."[14] Then he tried to describe the wave function of all the pairs with the help of an expression analogous to one he had come across in an article by Sin-Itiro Tomonaga about the "cloud" of unstable subatomic particles called *pions* around a proton or neutron at rest. Schrieffer's application was completely different, but it worked. He finished his calculation with a good wave function and returned happily to Urbana. Had Schrieffer been more up-to-date about the recent work of Bardeen's theoretical group, he would quickly have realized that the wave function he had just written down was not very different from the one that Lee, Low, and Pines had used to describe the polaron. As soon as Bardeen saw what Schrieffer had done, he understood that they had their solution.

The trio then spent thirteen days in frenzied labor calculating all the known properties of superconducting materials. As Bardeen tells the story: "In working out the properties of our simplified model and comparing with experimental results on real metals, we were continually amazed at the excellent agreement obtained. If there was serious discrepancy, it was usually found on rechecking that an error was made in the calculations. Everything fitted together neatly like the pieces of a jigsaw puzzle. . . . As we have seen, our model is exactly of the sort which should account for superconductivity according to London's ideas."[15] The group paid particular attention to the calculation of the relaxation time of the precession signal in nuclear magnetic resonance experiments.

> The nuclei of many atoms, including hydrogen and certain metals, have spins and magnetic moments, just like the electron. They too act like tiny magnets, and they will spin around the direction of an applied magnetic field the way a gyroscope spins around a vertical axis, the direction of the earth's gravitational field. In a nuclear magnetic resonance experiment, the tiny magnets are all set in motion in the same direction with the same speed, like a corps of ballerinas, and then left undisturbed. Little by little, some of them will slow down or speed up, and the corps of magnets becomes quite disordered. After a period known as the relaxation time, much of the original ordered motion has disappeared. The value of the relaxation time is important because it contains information about the mechanisms responsible for the loss of synchronism. In metals the relaxation depends on the interaction of the spin of the nucleus with the spin of the electron, which in turn depends on whether or not the electrons are superconducting.

All the two-fluid models predicted that this relaxation time must decrease with temperature. The BCS model, on the other hand, predicted that it would first increase and then fall sharply as the temperature continued to decrease. While they were finishing the calculation, the first experimental measurements showed they had it right.

There remained only the announcement of the long-sought result. At the last moment Bardeen asked his longtime colleague and friend Frederic Seitz whether, all deadlines having passed, their results might, nevertheless, be presented at the March meeting of the American Physical Society, the major annual meeting of condensed-matter physicists. Bardeen got his wish, but he didn't go to the meeting. So as not to diminish the luster of his young colleagues, he insisted that they alone present the results, while he stayed in Urbana. The complete article appeared in the *Physical Review* the same month.[16]

The BCS paper was one of those few fundamental papers that, from time to time, perturb the smooth, even routine, progression of scientific life by introducing concepts so new that a field is changed forever. Of course, this kind of general statement is ambiguous, because every scientific paper is supposed to present new information and original ideas that must be considered from then on. How to select from the hundreds of thousands of publications annually? There is no absolute criterion that measures the good and the less good, to the great chagrin of all those managers who would love to be able to arrange their researchers in order of merit, like horses finishing a race. Nevertheless, the three articles by Einstein that appeared in 1905 in *Zeitschrift für Physik*, the article by Watson and Crick published in 1953 in *Nature*, and the article by BCS (this is not an exhaustive list) have been so important to the history of ideas that one can, without too much risk of criticism, put them on a pedestal. In the library of our laboratory, of all the thick volumes of the *Physical Review*, the BCS article is the only one whose well-thumbed pages have fallen loose from their bindings.

The BCS theory is interesting not only because of its major contribution to fundamental science but also because of its absolutely unique role in the further history of superconductivity. In order to understand this better, we will try to review its major features as simply as possible.

A number of attempts have been made to describe the BCS mechanism, more or less graphically, more or less exactly, with a variety of analogies. Anglo-Saxon puritanism and French bawdiness will both permit a hearty chuckle over the idea of comparing the attraction between two electrons via the lattice to what happens to a couple sharing the conjugal bed. When one of the partners turns over to fall asleep, he leaves behind a depression in the mattress, which his partner immediately falls into. The most imaginative metaphor, which in our opinion explains nothing at all and certainly not what pairing is, was proposed in an ad by an English producer of cryogenic material. Superconductivity was compared to a couple of old cars on a badly paved street that are continually colliding to equalize their kinetic energies. At the critical temperature the energy gap necessary to break up the interaction between the cars is reached, and the cars become racers at the Indy 500 with neither friction nor resistance. The naïveté of some of these metaphors does not necessarily mean that they were intended for a popular audience. Until 1986 no one outside the small community of specialists cared one whit about superconductivity. These simplistic representations were, above all, attempts to reach other physicists who wanted to learn something about the new theory of superconductivity.

The most telling of all the descriptions we have found is the one proposed by Victor Weisskopf.[17] It appeared as one of a series of luminous articles in which he revisited numerous aspects of modern physics in clear language free of pedantry and accessible to a (very good) senior in high school.

> Weisskopf talks about an idealized metal, a lattice of ions with electrons circulating freely. Because electrons obey the Pauli principle, the only electrons that can interact with one another are those that have energies near the Fermi energy. Weisskopf's ions are large balls that vibrate around their equilibrium position; the frequency of their vibration can be easily calculated. When an electron with its negative charge traverses the lattice of large positively charged balls, it attracts them, especially those near its path, and leaves in its wake ions that have been slightly displaced and take some time to return to their equilibrium positions. With simple arguments Weisskopf calculates the length and diameter of this wake, which he represents as a tube in which the positive ions are a little more concentrated than elsewhere. During its brief existence this slight excess positive charge

attracts another electron, so that one could say that the first electron attracts the second via the intermediary of the moving ions. Simple arguments from classical mechanics suggest the order of magnitude of this attractive force; they give about the same answer as the most rigorous quantum mechanical calculations. Weisskopf completes his description by showing that the only electrons that are affected by the attractive force are those that move in a direction opposite to that of the electron that left the wake. The strength of the interaction depends on the volume of the tube, that is, on the distance over which the ions are displaced and the time it takes for them to get back to their positions. It's easy to understand that the volume of the tube will depend also on the mass of the ions, as Weisskopf's calculations show. When all is said and done, he provides a simple qualitative and quantitative picture that contains the essence of a Cooper pair.

A Cooper pair is an object that is totally unusual in traditional quantum mechanics, but its originality is no easier to perceive for a lay reader than the subtlety of an attack by a grand master would be for a beginner in chess. The fact that two objects that attract each other form what is called a bound state is in itself not surprising, and the objects don't even have to be quantized. It is sufficient to try it with two magnets. The strange thing about Cooper pairs lies in the fact that the two partners in the bound pair, the two electrons coupled by Fröhlich's interaction, will repel each other if they get close enough, even though they are moving relative to each other with dizzying speeds, of the order of several tens of thousands of kilometers per second. These features make it difficult to get a good intuitive idea of how the pair can form and still harder to decide what its properties might be. Even Weisskopf did not succeed in doing this; he had to have recourse to a little quantum mechanical calculation, which he made as simple and intuitive as possible but still managed to cover two pages with algebra.

This difficulty in describing Cooper pairs in elementary terms may explain why it took so long to discover them. Feynman had recognized this problem clearly during the Seattle conference, when he recounted his efforts to improve Fröhlich and Bardeen's model:

I always tried to do it the way Fröhlich and Bardeen did, that is, to solve their model with a precision better than the difference in energy that was to be demonstrated. I made diagrams and loops [Feynman's specialty]. . . . I calculated the specific heat with enormous precision but it was always proportional to the absolute temperature [i.e., just like a normal metal]. However, one of my students, Michael Cohen, remarked that the series that I was calculating did not lead to the lowest possible energy. Take the case of a harmonic oscillator (a pendulum) and consider the effect of a small perturbation εx^3 on the ground state. No matter how small ε is, one can never obtain the response by perturbation theory even if the series is perfectly convergent. . . . In other words, what's happening is that one is starting with a function that is qualitatively different from the right function so the right response is gone from the start.[18]

A calculational technique often used in quantum mechanics is based on the principle that small causes often have small effects; this is *perturbation theory*. As its name indicates, it consists in perturbing the system by adding a small constraint. If all goes well, the system reacts to the constraint by changing slightly; the smaller the perturbation, the smaller the change. Simple and well-tuned mathematical tools exist to determine what the effects of the perturbation will be by making use of the ingredients of the unperturbed situation and changing them slightly, just enough to let the perturbation express itself. This method has been tested over more than thirty years in all sorts of fields, including atomic physics, nuclear physics, and chemistry. The problem is that Cooper pairs cannot be constructed by treating the Fröhlich interaction as a perturbation of a Bloch wave, even though these waves had described very well most of the properties of metals. This is probably what explains the thirty years of failure that the theorists had had to put up with; they thought they had tried everything.

Constructing a Cooper pair requires two electrons with energies near the Fermi energy, with speeds in opposite directions and spins opposite as well. In addition, they have to be attracted to each other, but this force can be quite weak. (The Fröhlich interaction will work, but it is not specific, and any other force of the same type will do the trick.) Under these conditions two electrons at the top of the Fermi sea will form a bound state whose energy is slightly less than twice the Fermi energy; the bound state covers a region that is thousands of times larger than the distance between two neighboring atoms. If you remember London's program, you will note that a Cooper pair certainly looks like a good candidate for making a superconducting state, because its large spatial dimensions give it a macroscopic character and because it takes a nonzero energy to break it apart into two ordinary electrons.

The properties of a Cooper pair illustrate well the unique character of this object. In fact, its energy is lower than twice the Fermi energy of a single electron. That's why it is stable. Normally this is impossible, because the Pauli principle dictates that all states with energies lower than the Fermi energy are occupied at zero temperature. In fact, the pair is a new quantum state whose existence is a complete departure from everything that had been known before. The spatial extension of the pair is equally a completely novel property, which the classical image we have described here does not explain. With Fröhlich's interaction, its size is much larger than the wake of a single electron. Moreover, the pair extends in all directions, like a sphere, even though a classical trajectory would normally be a straight line.

Starting from this point, the work that had to be done was not trivial. Because the pair could not be constructed by slightly perturbing the normal states of the metal, it was necessary to construct a new state out of pairs of electrons, making sure that the two electrons were moving in opposite directions with opposite spins. Once this new wave function was constructed, everything that had been calculated for normal metals over thirty years had to be recalculated, without forgetting to introduce into the calculations the electron-phonon interaction that was indispensable. The task was not easy, because new rules of the game had to be developed to manipulate this unorthodox wave function. This was Schrieffer's contribution.

And so it is that the interaction between electrons via the intermediary of the vibrations of the ions allows the electrons to be associated in pairs. These pairs form a condensed state within the metal, whose energy is lower than the energy of the normal metal when the temperature drops below the critical temperature for superconductivity. This is the superconducting transition.

The article that Bardeen, Cooper, and Schrieffer sent off to *Physical Review* was a veritable compendium, in which not only was the model described in detail but many of its applications as well; all the known properties of superconductors were calculated and explained. This one article included the material of a dozen of today's papers. (In 1957 the reputation of a scientist was not based on the number of papers he had published.) Fifteen years later they received the Nobel Prize.

John Bardeen's long pursuit ended in triumph — the fundamental puzzle of superconductivity was solved. The program as it had been defined by London and recapitulated by Feynman had been accomplished. This brilliant tour de force constitutes a good example of traditional scientific method: a spectacular simplification, what the philosophers call a "reduction," led to the concept of pairing, which in turn permitted BCS to furnish for the first time a unified description of the behavior of a superconductor. A new era had opened for superconductivity.

CHAPTER 17

The Golden Age

The article by Bardeen, Cooper, and Schrieffer—collectively known as BCS—was hard to digest. So many problems were discussed, so heavy was the mathematics, so difficult were the presentation and calculational techniques, that most readers found the paper burdensome. As P. W. Anderson has remarked, the fact that the article was so poorly written was really fortunate for other physicists. It gave them a chance to publish their own articles covering the same ground in language that was more accessible and with mathematics that might be more easily adapted to new problems. Even though many phenomena were satisfactorily described in the original article, a deep understanding of the theory was reserved for a small number of initiates.

Nicolai Bogoliubov, a Soviet physicist almost completely unknown in the West, was one of the first; he led a group of theorists in charge of developing bombs in the Soviet atomic center. For a dozen years he had been developing an elegant formalism that would enable BCS theory to be rewritten from a new point of view. We have met Bogoliubov earlier, at his 1946 seminar at the Institute for Physical Problems, where he presented his microscopic theory of superfluidity in helium, which was coolly received by Landau's group.

Benefiting from the East-West thaw of 1959, Bogoliubov stepped to the forefront as far as the West was concerned when he published his work in the Italian theoretical journal *Il Nuovo Cimento*. This was an elegant and concise reformulation of BCS theory that would make it accessible well outside its original domain.

> The main advantage of Bogoliubov's method was that it translated BCS theory into the language of self-consistent fields, or Hartree-Fock theory, which had been used since the beginning of quantum mechanics. The principle is simple and can be summarized in a brief dialog: "Suppose the problem is solved," I say, and you respond, "Then let's see if it works." In other words, suppose you know the form of the solution, the wave function

that describes the superconducting electrons, not just for infinite homogeneous metals with no external field (a "trivial" BCS solution) but for all sorts of cases. The crystal might be finite, with various faults and dislocations, and the ions magnetic. The wave function would be expressed in terms of some parameters with unknown values that can be determined by using this trial wave function to calculate the energy of the system. In this calculation you include all the specific elements of the system — the faults, impurities, and boundaries, for example, but also the effective interaction of the electron with the other electrons, all represented by the same trial wave function. You carry out the calculation with this wave function and interaction, and you decide whether the energy is indeed a minimum. If it is, you quit; you've found the right values of the parameters. If not, the usual case, you change the parameters until you've found the minimum. Nowadays, with big computers, this task is not so difficult, but this was not the case forty years ago. When you've found the minimum energy, you've found the self-consistent solution — the potential energy (the interaction) determines the wave function that fixes the positions that determine the potential energy; the cat eats its own tail.

This method, applied by Bogoliubov to superconductors, yields a system of coupled equations as a self-consistency condition. Quantities that appear in one equation are determined by solutions of the other equation, like two equations with two unknowns in algebra. The equations look like Schrödinger equations with an extra term, called the *pair potential*. The magnitude of this potential at a particular position reflects the condensation of pairs at that position; it is one of the parameters determined by a self-consistent solution. By solving the coupled equations of Bogoliubov, the Russian school was able to demonstrate that the pair potential is the solution to what are none other than the Ginzburg-Landau equations.

Landau's group did not let Bogoliubov maintain a monopoly on alternative formulations of BCS theory. In 1959–1960 L. P. Gorkov, a young theorist at the Institute for Physical Problems, presented BCS theory in the remarkably dense language of quantum field theory, which had been little used by condensed-matter physicists. Landau's school would make abundant use of Gorkov's ideas in trying to understand systems that had hardly been accessible — alloys, for example, or superconductors with magnetic ions.

At this stage it may be useful to stop for a moment and step back a bit from the story we're telling. In the last few paragraphs, with little warning, we've spoken about physical theories as languages. In fact, different theories can describe one reality with different equations and vocabulary; they seem to be different languages. This is in part because there is no way that we, in this book, can discuss the technical details of calculations that are obviously understandable only by specialists. Nevertheless, one fact is inescapable, for specialists and nonspecialists alike: the same physical phenomenon can be accommodated in different descriptions that are rigorously equivalent even if they are conceptually distinct. This equivalence is not a given; it must be rigorously proved. Sometimes the proof itself requires that new tools be created to enable easy transfer from one conceptual framework to the other. Physical theories do not progress only by enlarging the scope of previous theories, as one might naively think, but also by passing from one kind of language to another. From this point of view the work of Bogoliubov and Gorkov can be considered as having provided tools that led to a microscopic justification of the phenomenological description of superconductivity provided by Ginzburg and Landau.

Their methods were applied most systematically in France, with the formation of the superconductivity group at the University of Paris in Orsay, just outside the city. Pierre-Gilles de Gennes, a theorist who received the Nobel Prize in Physics in 1991 for other work, was the group leader. While it's easy for us to think of many prestigious physicists from the era around 1960, we have to admit that physics in France at that time didn't amount to much. The war was one reason, but the situation of the French universities was another, and it predated the war. Between the two world wars, at the Ecole Normale Supérieure in Paris only two students per year could enter the physics program. One had a scholarship to write a research thesis and the other to prepare for the "agrégation," a special degree for a very small percentage of high school teachers who perform superbly on a rigorous set of exams. (According to Jacques Friedel, this examination "killed research between the two wars.")[1] The situation at the Ecole Polytechnique in Paris was no better. Only a few branches of French physics were really vigorous in those days. One was experimental nuclear physics led by Irene and Frederic Joliot-Curie; there was physical chemistry, with Paul Langevin, Jean Perrin, and Edmond Bauer, and optics, with Jean Cabannes.

Before the arrival of the Nazis, Germany was the world center of physics. French physicists had been one step removed since the war of 1870; Edmond Bauer and Léon Brillouin were two exceptions. Bauer had gone to Strasbourg, near the border with Germany, for historical reasons the only university in France with a reputation in research comparable to its peers in Germany. He taught statistical mechanics, and it was he who had tried to keep Fritz London in France. Brillouin had worked with Sommerfeld in Munich and taught at the Collège de France. He could well have established a school, but he could never attract thesis students because of his difficult personality. So, nothing here to give birth to a French tradition in condensed-matter physics and still less to a team of specialists in superconductivity. Cambridge, by comparison, in a country where conditions were scarcely more favorable, managed to graduate twenty to forty thesis students per year. In the British system these graduates could go on to teach in secondary schools without having to pass hurdles anywhere near as difficult as the agrégation.

Low-temperature physics in France really got started again in Grenoble after World War II. Louis Weil and Louis Néel took over the grand tradition of magnetism that had begun with Pierre Curie, developing and extending it in the direction of very low temperatures, down to that of liquid helium. The French company L'Air Liquide, which had for a long time supported the research that it needed at the Institut Fourier in Grenoble, created its own laboratory in Sassenage, in the same region. Thus, there was a collaboration between the university, the CNRS researchers, and industry for solving technical problems in cryogenics, yet it never carried over into what we would really call physics, in spite of the strength of the L'Air Liquide group. Always on the lookout for new ideas and ready to recruit outstanding young Englishmen, Weil was responsible for having B. B. Goodman come to Grenoble after the BCS publication. With the help of strong unions, the salaries of French government researchers easily withstood comparison with those in England at this time, when research entered a new era. It was Goodman in the West who suggested that type II superconductors have a regular magnetic structure, layers that the magnetic field might penetrate. With a rare sense of fair play, he gave credit to Abrikosov for his correct description in terms of a regular network of vortices. Goodman later returned to England, to the British Oxygen Company, where the remuneration (in industry), without unions but with the same Keynesian expansion as in France, largely surpassed that of the government researchers in the CNRS.

It was around this time that de Gennes completed his work with Charles Kittel in Berkeley[2] and rejoined the condensed-matter physics laboratory created in the School of Science at Orsay by Raymond Castaing, Jacques Friedel, and André Guinier. Having learned about the work of Ginzburg and Landau, de Gennes definitively decided in 1961 that he would specialize in superconductivity. He first started a theoretical group and then a group of experimentalists, all branded with his own trademark style.

For several years, then, in the early 1960s French researchers found themselves for the first time at the cutting edge of research in superconductivity. The theorists demonstrated all that could be learned from a thorough application of Ginzburg and Landau's formalism. The experimentalists attacked a large number of complex situations with remarkably simple techniques: the contacts between two metals, one normal and the other superconducting; amorphous superconductors; magnetic impurities; superheated superconductors; surface superconductivity, and others. These issues all involved inhomogeneities in superconductors.

Physicists working on superconductivity through the 1950s had not systematically considered the magnetic structure of their materials, which might give rise to spatial variations of their superconducting properties, or even about the fact that their specimens were not homogeneous and infinite in all directions. The Ginzburg-Landau model opened the door to the understanding of inhomogeneous materials, and BCS could provide a description of transport properties. A marriage of the two might be very fruitful, de Gennes thought, particularly for experimentalists. Spatial variations could arise from limits imposed by the dimensions of the specimen — in thin films, for example, or in the contact between an ordinary metal and a superconductor. They might originate in the internal structure of these specimens (inclusions or magnetic impurities) or finally in the inhomogeneities inherent in a superconductor in the presence of a magnetic field. The model of Bogoliubov was particularly well suited for this issue because the pair potential that it contained could introduce spatial variations. The Ginzburg-Landau equations, integrated into all possible or even imaginable situations, furnished an abundant catalog of spatial modulations of the order parameter. The close proximity between the experimentalists and the theorists gave the group of de Gennes a particular dynamism that was highly efficacious.

Saclay, Grenoble, Paris, and other smaller centers also contributed, sometimes with extremely sophisticated experiments, like the discovery of the vortex structure of type II superconductors by low-energy neutron diffraction, neutron scattering that emphasizes the wave nature of neutrons.[3] Within a few years more than fifty scientists in France were involved in superconductivity. France was now a force in the international superconductivity community, but that meant nothing as far as an industrial research policy was concerned. In fact, the small world of superconductivity in France summed up in itself many of the virtues and shortcomings of research *à la française*.

The French renaissance in superconductivity was not exceptional. With the East-West détente and economic prosperity, scientists took up traveling and international exchanges once more. Germany, England, the Netherlands (and not just Leiden), Italy, Spain, and other countries had groups engaged in superconductivity. Important experimental results might come one day from one country, the next day from another, and so nourish the progress of the field.

A Scandinavian engineer who became a physicist, Ivar Giaever, provides an example. He decided to study the behavior of tunnel junctions, a kind of sandwich in which a very thin insulating layer, maybe a few tens of angstroms thick, separates a superconducting electrode (in his case) from a normal electrode. When a voltage difference was applied to these electrodes, Giaver observed that above a certain threshold a current flowed from one electrode to the other in spite of the insulating layer, and it varied strongly with the applied voltage.[4] The magnitude of the voltage necessary turned out to be a direct determination of the energy gap in the superconductor.

The possibility of *tunneling* had been known since the early days of quantum mechanics; it is a uniquely quantum mechanical effect. A sled starting from rest at the top of a hill and picking up speed as it goes downhill cannot possibly get over the top of a nearby hill higher than where it started without a push. The equivalent process, in which a particle can get from one side to the other of a potential hill that seems too high, is allowed by quantum mechanics at the atomic scale, but the particle goes through the hill instead of over the top. Alpha decay of nuclei was the first experimental demonstration: the alpha particle escapes through a "hill" of potential energy too high for a classical particle. We say that it "tunnels" through the potential barrier. With his experimental setup Giaever transposed this idea of a barrier into something very concrete. The thin insulator in the tunnel junction, a layer of oxide, *is* the potential barrier for electrons wanting to pass from one electrode to the other. The superconducting gap had been previously accessible only with great difficulty, by infrared reflection, for example, ultrasonic absorption, or surface impedance. All at once it became measurable via a modest but delicate oxidation of a metal surface followed by deposition of another superconducting metallic layer on the oxide barrier. This pioneering experiment won a Nobel Prize for Giaever. The same principle has now been applied in many areas, and it is the basis for the extensively used scanning tunneling microscopes that allow individual surface atoms to be observed.

Tunneling experiments also yielded the strength of the electron-phonon interaction in superconductors; the calculations that determined the strength were considered sensationally precise at the time. A clever analysis of the tunneling effect led to the discovery of one of the most surprising aspects of superconductivity, the Josephson effect. It was first described in 1962, in a brief and rather obscure article in *Physics Letters* by Pippard's student Brian Josephson, who was then twenty-four years old.[5] What is remarkable about this article, as noted right away by Philip Anderson, who was giving a series of summer lectures in Cambridge at the time, is that the predicted phenomenon is described in its entirety, in all its aspects, with all its expected consequences. Patent hunters simply had to throw in the towel, even though the Josephson effect — or, more precisely, effects — are still the best source of applications of superconductivity.

BCS theory took account of the electron-phonon interaction by a phenomenological term that the standard models used to describe solids could not determine. G. M. Eliashberg, a Soviet physicist, by careful use of field theory (once again) proposed a model that allowed this term to be predicted once the electronic structure of the metal was known precisely. J. M. Rowell and W. L. McMillan transformed these results into an algorithm that has been widely used by specialists in the tunneling effect in superconductors.

The Josephson effect is not easy to understand and even less easy to explain in intuitive fashion, but the basic experimental facts are straightforward to describe. A current of Cooper pairs, a supercurrent, can flow between two superconductors separated by a thin layer of insulator even when there is no voltage difference between the two superconductors. This is the direct current (DC) Josephson effect. If, instead, a voltage difference exists across the boundary, then there will be an alternating current between the two superconductors. This is the alternating current (AC) Josephson effect. The two types of current both depend on the relative phases of the wave functions in the two superconductors, and that's where the problems in understanding arise. No doubt it is for this reason that ten years elapsed between the BCS paper and the Josephson paper, even though effects that are apparently much more complex were explored, until the well went dry during this period.

The absolute phase of a wave function seems to have no physical significance whatsoever. That's because the absolute phase is an angle whose value doesn't appear in the calculation of any physically measurable observable, such as the position or the energy of an electron. The relative phase between two different wave functions does matter, however, and gives rise to interference phenomena.

The word *interference* is used by analogy with Thomas Young's famous two-slit experiment in optics that demonstrated the phenomenon of optical interference. In Young's experiment light waves from a single source pass through two narrow slits side by side in order to reach a screen. The light rays are diffracted, bent, at the slits, and the light rays reaching a given point on the screen have traveled different distances. Thus, the light wave from one slit at this point on the screen may be at its peak, and the other at its valley; if so, they will cancel out and give a dark spot on the screen at this point. The two waves are said to be out of phase; their phase difference is then 180°, π radians. If both are at their peak at a given point, the two will add, and there will be a bright spot. The two waves are in phase, and the phase difference is zero (or 2π, 4π, etc.) Thus, a succession of light and dark lines, called *interference fringes*, is observed.

The Josephson effect also depends on a difference of two phases, that is, a relative phase, so that aspect was not new. Normally, however, with *microscopic* objects, the phase of a wave function changes with position, and so the phase difference would change with position as well. The Josephson effect, on the other hand, depends on the relative phase between two *macroscopic* objects, and, for superconductors at least, the phase is essentially the same everywhere in the material. This is where the problems arose.

The idea that the superconducting wave function is rigid and macroscopic, that it varies little from one point to another of the superconductor, seemed contradictory to the existence of supercurrents. The supercurrent flows in a certain direction with a well-defined magnitude in a given situation; this would indicate some kind of mechanism that lures the electrons in that direction. For Fritz London the cause of this motion lay in the fact that the phase of the wave function varied from one point to another. He had even demonstrated that the supercurrent at each point was proportional to the rate of change (the gradient)[6] of the phase at that point, and it was in the direction that the phase varied most quickly. This result was not in contradiction with traditional ideas, because it involved only the difference between two phases, and not their absolute value. Nevertheless, the phase has one and only one value at a given point. The fact that the phase had to return to this value after traversal of a complete circuit had led London to predict the quantization of magnetic flux. This spectacular effect had just been verified almost simultaneously by R. Doll and M. Näbauer in Germany and by B. S. Deaver and W. M. Fairbank in California.[7] (Fairbank knew London well when Josephson wrote his article; they were at Duke University together.)

The *phase* of a wave function is not simply a property of a wave with purely mathematical significance. It plays an important role in maintaining what physicists call gauge invariance in quantum mechanics. The magnetic field is a physical quantity represented at every point in space, inside a material as well as outside, by a vector (think of an arrow) **B**. The length of the vector represents the strength of the field (what we've labeled **B**) and is given in units of teslas or gauss, depending on whether one is an engineer or a physicist (1 tesla $= 10^4$ gauss). The vector (arrow) points in the direction of the field. Unfortunately, this vector magnetic field is a difficult object to manipulate mathematically. Although it has been known since the time of Ampère (1820) that there is a close relationship between magnetic fields and electrical currents, the mathematical expression that links the two is somewhat cumbersome. (The reader may remember various rules with thumbs, fingers, and hands for figuring out the direction of one relative to the other.) Thus, physicists were led by Maxwell to use another vector in talking about magnetic fields, the vector potential, **A**. **A** is more manageable than **B**, and its relation with the current is easier to calculate. In addition, the expression for magnetic energy can be written very simply in classical electromagnetism in terms of **A**. It's for this reason that quantum mechanics has usually chosen the vector potential rather than the magnetic field **B** to introduce magnetic effects into Schrödinger's equation and into the Ginzburg-Landau equations as well. There is one problem with **A**, however, which can prove to be a real nuisance sometimes: there are an infinite number of vector potentials, all of which correspond to the same magnetic field. This looks damning, but, before swallowing the cyanide, let us suggest a different pharmaceutical remedy, a comparison.

The active ingredient in a medicine is most often a particular molecule that has to make contact with one or another organ inside the body. Now there are all sorts of ways that the molecule might pass from outside to inside — a pill maybe, gel, syrup, suppository, injection, and so forth. It's always the same molecule but presented in a variety of forms. The body functions as a filter that eliminates all inessential wrappings and keeps only the crucial core. This is a bit like the vector potential. The mathematical operation that takes you from the vector potential to the magnetic field must clear away all the frills and provide only the magnetic field strength, which is the physically measurable quantity. When the vector potential appears in an equation that is supposed to provide a physically measurable quantity, it is necessary that a compensating mechanism be included in the equation that will guarantee that the value calculated is the same regardless of the choice of vector potential. This is precisely the role of the phase of the wave function: it is there to make up for the variations in the choice of vector potential. It is said to "maintain gauge invariance."

Thus, the existence of the phase was justified in the eyes of some physicists by the role it plays in maintaining gauge invariance. No one realized how useful it would be for Josephson. His idea, however, was quite simple. When you take two distinct superconductors that don't touch each other, the superconducting wave functions that describe the paired electrons in the two samples are completely independent. In other words, the phase of one is random relative to the other. Josephson assumed that, if the two samples got close enough together, the two wave functions would cross the barrier between them and couple to each other as they would in the tunneling of electrons between ordinary conductors. The coupling would be weak but real, and, in particular, it would "lock" the phase of one relative to the phase of the other. In these conditions, said Josephson, because the wave functions of the two samples partly overlap, there's no reason why a supercurrent, a current of electron pairs, could not flow from one sample to the other. The current would certainly be small, much smaller than the intense currents that can flow in a standard superconducting sample, but a supercurrent nonetheless, a pair current that needs no voltage difference in order to flow.

The sign of the phase difference determines the current direction; the intensity of the current depends on the magnitude of the phase difference. The Josephson current is zero when the phase difference is zero; thus, it must also be zero when the phase difference is 360° (2π radians), 720° (4π radians), and so forth. If the phase difference is not zero, the current will flow toward the higher phase, exactly as it does in an ordinary superconductor. Josephson showed that the proper mathematical function, which ensured that the current would be the same each time the phase difference increased by 2π, was a sinusoidal function. He thus proposed that the current I traversing the junction followed the expression $I = I_0 \sin(\phi_2 - \phi_1)$, where I_0 was determined by the specific type of junction and ϕ_2 and ϕ_1 are the phases of the two superconductors at the junction. The junction itself is now called a *Josephson*

The golden age in Cambridge: the Mond Laboratory in 1963. (1) David Shoenberg; (2) Brian Pippard; (3) Brian Josephson.

junction. The junction might be a tunneling junction, in which the two superconductors are connected by a thin insulating layer, such as an oxide, or it might be a weak junction, in which they are connected by a thin layer of normal metal.

Thus did the phase difference, whose value entered explicitly into the expression of a measurable physical quantity, become here in some sense a physical object, just like the phase difference of light waves, which plays a fundamental role in interference experiments and holography. Nothing more was required for some insightful inventors to wonder whether it was possible to construct *quantum interferometers* based on the Josephson current as a measure of the phase difference between two circuits. In optics it's easy to introduce a phase difference between two light waves: it suffices to make the light waves travel different distances. Could the same result be obtained with quantum waves? The tight link between the phase and the vector potential imposed by gauge invariance can do it. By varying the flux contained in a closed circuit in the shape of a ring, the vector potential varies around the ring, and so does the order parameter. Cut the ring in half, into two semicircles that are almost touching, and you have a ring with two Josephson junctions. This is the basis of a quantum interferometer. In fact, such an apparatus constitutes one of the first applications of the Josephson effect, and it has become the most sensitive detector of magnetic fields ever constructed. Other applications of superconductivity have not necessarily been so successful.

CHAPTER 18

After the Golden Age
Tomorrow, Always Tomorrow

After the excitement surrounding the triumph of BCS, the pure intellectual curiosity aroused by superconductivity slowly gave way to discussions of usefulness. The quest for applications began to replace science. The world of superconductivity, like all of physics, had grown from year to year and was now part of big science, even if it represented only a small fraction. Research goals no longer had to be justified to a mere few hundred specialists with, officially at least, a common taste for fundamental ideas. A community of several thousand researchers was often engaged in simply duplicating or paraphrasing the work of others, an effort justified only by its relevance to potential applications.

This period, post-BCS, furnished the all-purpose sales pitch for innumerable grant proposals requesting funding for research claiming to lead to applications. Government sponsors wanted to hear such arguments; so did the few firms whose government contracts enabled them to work in the area. The economy was in full Keynesian expansion. Even the Vietnam War did not weigh too heavily, at least in the beginning, on U.S. research budgets.

Applications of superconductors had been conceived right from the start — superconducting coils, for example, for producing very intense magnetic fields — but they were hardly feasible at the time. Little was known about the behavior of different superconducting materials in high magnetic fields or with high currents. Given the cost of experiments, it was out of the question to set about a program of pure empirical exploration of the properties of superconducting wires as had been traditional in the electrical industry (a tradition that is part of what one discreetly calls the "art of the engineer").[1] The problem was clarified considerably when the distinction between type I superconductors (Pippard's, like lead) and type II superconductors, which have much higher critical magnetic fields, was understood.

To reach high fields you definitely had to use a type II superconductor, and not the lead that Kamerlingh Onnes tried. These type II materials had the advantage of high critical temperatures, about 13K, well above the 4.2K of liquid helium. The only problem was that techniques for making wires out of niobium or its alloys were

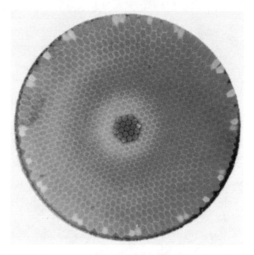

A superconducting wire. This is a cross-section of a superconducting wire produced by the French company Alsthom-Belfort for alternating current. The diameter is 0.3 mm, and the wire includes almost one million filaments of niobium-titanium (NbTi) twisted in a complete circle in steps of 1.5 mm. The filaments, which represent about 20 percent of the mass of the cable, are supported in a matrix made of copper and a copper-nickel alloy (the hexagons visible in the figure). At the (European) standard household frequency of 50 Hz, the wire can be used in magnetic fields up to 1 tesla.

not well known; they were brittle and stiff, more like mattress springs than malleable conductors. Once this problem was solved, another arose right away. If part of a wire accidentally went normal, because of a bad contact, for example, all the energy stored in the windings might be suddenly released. The whole type II superconducting wire could become resistive because, as we have seen, the mean free path of electrons is small. The scenario could be catastrophic. The cryostat could explode, since helium requires just a small amount of energy to vaporize and multiply its volume by a factor of seven hundred. Even if this worst case did not occur, the most likely outcome would be bad enough, a permanently warped coil and local shorts in the windings.

To make these transitions less abrupt, composite wires had to be developed with thousands of superconducting filaments embedded in a copper matrix. In normal use the superconducting filaments short-circuit the copper; the wire acts just like a pure superconductor. If it tries to go normal, however, the superconducting filaments become highly resistive. The electric current finds a better path through the copper, which has little resistance, so the whole wire heats up more slowly and ultimately doesn't get very hot. Perfecting the different steps of this solution (choice of superconductor, metallurgy of the chosen material, techniques for extruding the composite wire, optimization of the windings, construction of reliable and user-friendly cryostats) required a number of years and the help of specialists who a priori had nothing to do with superconductivity. To take one example, it was not an easy job to convince workers and managers in the extrusion shops, whose techniques had not changed since the advent of the electric motor, to get used to working in a dust-free environment.

Be that as it may, understanding the nature of type II superconductors and how they behaved in a magnetic field, and the subsequent development of multifilament superconductors, led to a broad range of applications of superconductivity labeled "high current." The first is the one most easily understood by the public: if the electrical resistance is rigorously zero, it's possible with a superconducting cable of modest cross-section to carry electric currents of hundreds of thousands of amperes without losses. Cryogenic cables of this sort were constructed for all sorts of superconducting electromagnets: solenoids for creating magnetic fields more than a hundred thousand times the magnetic field of the earth, coils of huge dimensions for high-energy physics and, more recently, for sensitive medical examination of the human body, as in magnetic resonance imaging devices (MRI). In 1963, for the particle accelerator at Fermilab near Chicago, four hundred superconducting coils were built, not without difficulty, and placed along an evacuated beam transport line to control the trajectory of protons accelerated to very high speeds. Such coils have since become almost commonplace. The idea of using enormous coils to store temporarily electrical energy generated when usage is low for release during peak hours has also been considered.

The high-current applications that have the most impact on the public, however, are related to transportation, such as trains suspended magnetically. The magnetic field that does the levitating is furnished by coils in the engine and in each car. The Japanese constructed a special experimental track for magnetically levitated trains. Speedboats with magnetohydrodynamic propulsion are accelerated by the force on a current-carrying conductor at right angles to a magnetic field. The current is in the water between two solid electrodes in the hull, and the field is created by a superconducting coil placed on board in the appropriate direction. Prototypes were made several years ago in Japan and in the United States; some people say that these are the boats of tomorrow.

In fact, even if popular articles about superconductivity always speak about levitating trains and lossless transmission lines, nothing of the sort is actually in service. Cryogenic transmission lines require that the cable be placed in a low-temperature channel permanently kept cold. The economic advantage is, however, far from obvious if the calculation of the costs of exchanging a classical transmission line for a superconducting one includes the write-off of the standard line withdrawn from service before the end of its useful life. The change in technique also involves losing the expertise of the maintenance crew. In fact, when traditional methods and innovative methods compete, it is often the traditional technique that carries the day at the price of continual small improvements.

This is what has happened with transmission lines. To reduce losses, transmission

Applications of superconductors.
Top: A Japanese train suspended by superconducting magnetic levitation.
Bottom: *Jupiter II*, a boat propelled by superconducting motors.

voltages have been increased little by little. Large quantities of energy are now transported with losses of only a few percent by very high voltage (million volt) transmission lines. The idea of lossless transmission by superconducting lines remains a slogan. Indeed, there would be absolutely no losses in the cables themselves if the current were continuous, but there will always be energy expended to keep the whole network cold. The advocates of superconducting lines have always agreed that, yes, helium is costly, but just wait until we can get superconductors at liquid nitrogen temperatures (77K). Well, now we have them. Yet, even if they could be put into service and perform right away as well as the best conventional superconductors, there would be no economic advantage to making the change.[2] If superconductivity could be achieved at room temperature, these statements would have to be reexamined.

The situation for levitating trains is a good illustration of the blindness manifested by the specialists when it comes to innovation. The Japanese managed to operate a magnetically levitated train with conventional superconducting magnets in 1979;[3] it carried no passengers but reached a speed of 517 km/hr. The jacket of a 1988 book on superconductivity assures us, apropos of the discovery of the first high-temperature superconductors in 1987, that "no other scientific discovery has the promise of such a great social, political and economic impact on the world."[4] It's a good example of this blindness. The authors, graduates of the best American universities, specialists in superconducting electronics, coolly (the least one can do for superconductivity) announced that: "Techniques for railroad traction are now at an impasse. Although modern high speed trains transport passengers at 185 km/hr, the engineers doubt that greater speeds can be attained without a fundamental change in the nature of the railroad."[5] Not two years later the TGV-Atlantic in France reached speeds higher than 500 km/hr, and the TGV runs regularly at more than 300 km/hr. The solution "iron on iron" (wheels on a track) is not exactly at an impasse. Our "specialists" can console themselves, however; this technical feat has profited from the world's most sophisticated metallurgy — but not superconductivity.

The other domain for applications of superconductors involves very low currents. These are based on another discovery of the post-BCS era, the Josephson effects described in the last chapter; they represent the most direct demonstration of the macroscopic quantum character of the superconducting state. While the high-current applications of superconductivity are all inspired by well-established techniques and apparatus, applications based on the Josephson effect are entirely new.

Although it's quite easy to devise an experimental setup to demonstrate the Josephson effect (inserting a simple constriction in a superconducting wire will suffice), it's a different story when the goal is precise reproducibility, and the techniques employed vary with the utilization envisaged. A popular item in condensed-matter physics labs is an ultrasensitive magnetometer, a device for measuring a magnetic field with a precision better than one part in a billion of the earth's magnetic field.

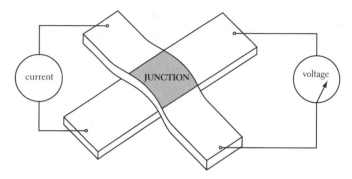

The basic elements of a Josephson junction. Two strips of superconducting metal overlap each other, with an insulating layer of oxide separating them at the crossing point. The oxide layer must be thin enough that Cooper pairs can tunnel through it. Simultaneous measurement of the current crossing the barrier and the potential difference between the wires, as shown, can determine the critical current of the junction — that is, the maximum current that can flow when the potential difference is zero. Placed in an external magnetic field of variable strength, the current in the junction is modulated, and very small changes in the field can be detected.

The basic element is a superconducting ring with a Josephson junction at one point. Without the junction a magnetic field in the hole would be trapped — the Meissner effect would not allow the field lines to cross the material of the ring, so the field in the hole would remain constant even if the external source field varied. Because of the junction, however, it's possible to vary the flux in the interior of this special ring by varying the external field. We know also that the magnetic flux in the hole in the ring is quantized, and so the flux changes only in steps of a single quantum, a very small unit. This accounts for the sensitivity of the device. The ring will react to an increase in the magnetic field applied parallel to its axis by increasing the superconducting current, to maintain the ring itself field free. If, however, this current exceeds the maximum current that the Josephson junction can support — the critical current — the junction will temporarily become resistive and the ring will no longer be entirely superconducting. A voltage difference then appears across the constriction, and the flux penetrates the ring until the current becomes lower than the critical value and the entire ring becomes superconducting again. Technically, this is done via a feedback loop that adjusts the current in the ring to keep the total flux through the ring constant. Each time the flux varies, the feedback loop increases the opposing current just enough to keep the total flux constant. The current in the feedback loop is then a very sensitive measure of the variation of the magnetic field.

By putting not one but two junctions in the ring, we get the original Josephson interferometer, a SQUID, which is not a mollusk with a lot of tentacles but a Superconducting Quantum Interferometric Device. Such an arrangement is sensitive to the phase differences between the wave functions that represent the superconducting state on the two sides of both junctions. By varying the flux inside the ring produced by an external magnet, the phase difference across one junction will vary relative to the phase difference across the other. The resulting current in the ring will vary sinusoidally with the magnetic flux; minima occur when the magnetic flux through the SQUID is equal to an integer multiple of h/2e (where e is the elementary electric charge and h is Planck's constant). The position of these minima can be used to measure very small values of the magnetic field, such as the extremely weak magnetic fields associated with biological activity.

In Josephson effect magnetometers that use only one junction, the junction is made by crushing the tip of a niobium contact point, like the one that makes the contacts in a lead sulfide crystal in old crystal radios. The complete apparatus was commercialized at the beginning of the 1970s by an American firm. Since then, devices have been fabricated that measure the projection of the magnetic field along all three directions or else one projection at a number of neighboring points. They can, for example, map the magnetic fields produced by brain activity. These magnetoencephalograms are now produced by devices containing hundreds of SQUIDS. SQUIDS are also used to detect submarines. In all these systems the two junctions must be essentially identical, so all the technology for making the thin films and microelectronics familiar in conventional electronics was soon transferred to superconductors. This also made it easier for Josephson junctions to be used as infrared detectors.

So far we have mostly discussed the DC Josephson effect. The AC Josephson effect is much more difficult to explain by simple arguments. It occurs when a constant potential difference is maintained across the junction. The junction then becomes a source of alternating current whose frequency is proportional to this applied voltage. The relation between the frequency n and the voltage V is universal; it does not depend on the junction: n = 2eV/h. This effect is reversible, which means that, if one puts an alternating current through the junction, a constant voltage appears across it. Infrared radiation is likely to induce such alternating currents; the ensuing voltage signal allows it to be detected.

A vast new arena for applications of superconductivity was opened in 1968, when IBM decided to seriously consider the feasibility of a superconducting computer. The program was so important for IBM that specialists in thin films and Josephson junctions were recruited from all over Europe and especially Scandinavia. After an initial salvo of articles, silence was the rule for some time; this reinforced the impression of real progress toward a working machine. You would search in vain the volumes of the scientific journals of IBM for the headings "superconductivity" or "Josephson" at the beginning of the 1970s.

The arguments that justified this effort are not difficult to understand. IBM wanted more and more powerful computers; the era of personal computers had not yet begun. A powerful computer is first and foremost a fast computer, so fast that the time for an electric signal to pass even the small distances between two points in a circuit is no longer negligible compared to the time constants of the electronics. Logical circuits had to be small and dense, so a powerful computer is necessarily a compact computer.

Now we have two reasons for superconductivity to enter the game. To be fast, the logical circuits must be fast. The lifetime of a Cooper pair, the time the two electrons of the pair stay linked to each other, is of the order of 10^{-12} seconds (a millionth of a millionth of a second). That's short! This lifetime fixes the scale of time for an elementary operation of the computer, the time for the passage from the state zero to the state one. To be compact, the different elementary circuits have to be piled one on top of the other, so there's little room for ventilation or refrigeration, and the circuits must dissipate almost no heat. The basic component of a superconducting computer would be a Josephson junction. The zero state corresponds to a current across the junction with no applied voltage. The value of this current is regulated so that a slight increase puts the junction into its resisitive regime; a voltage appears across the junction, and it's read as state one. The whole computer could be put into a shoebox, so a large cryostat wouldn't be needed.

Starting from these basic principles, the superconducting program at IBM gained some weight over the course of time. It soon became the second most important development program at the company, right behind gallium arsenide transistors, the fast transistors that arose from the evolution of the older techniques of semiconductors. After several years of development IBM slowly began to lift the veil of secrecy surrounding the results of its endeavors. The first demonstration of feasibility was announced in 1975: a number multiplier had been constructed entirely out of Josephson junctions, with a speed better than that of any equivalent circuits existing at the time. As a by-product, the company had also engineered the development of the entire cryogenic system for the computer. It looked like a desk of conventional dimensions, with room for a set of drawers on each side of the desk chair. Instead of the drawers, however, on one side was the cryostat that housed the computer, and on the other was a small liquefier continuously feeding the cryostat with liquid helium in a closed loop. A public relations campaign began in 1980, and with it came details of the circuits themselves.[6] Each electronics card in the shoebox would include tens of thousands of oxide junctions whose thickness would be regulated to a precision of one atomic layer. Sample logical signals demonstrated the enormous switching speeds possible. The response was not long in coming. A Japanese company decided to launch a program immediately with similar objectives; forty scientists were needed to get started. Even in France researchers in Grenoble at the Atomic Energy Commission were finally given the green light to work on Josephson junctions to use as the basic logical element of a new family of computers.[7]

The outlook was bright, or so it seemed. At the same time as these innovations were being revealed to the scientific public, visitors to IBM central headquarters in Yorktown Heights, on the banks of the Hudson in Westchester County, just north of

New York City, were reporting contradictory rumors. The program was losing importance; the principal scientists were beginning to publish interesting articles about the results of the program, but they were not directly oriented toward actually making a superconducting computer. A new manager had been named to head the two competing programs, superconductors and gallium arsenide, with orders to decide which one seemed most promising. It was the superconductor project, the more risky one, the one that represented a complete break with the past, that was abandoned at the end of 1983 for reasons that were never made public.

The arguments put forward to promote the project were now turned around to emphasize how unrealistic it was. It would take more than a hundred distinct operations to fabricate each junction card. The rate of rejecting faulty cards had to be less than one in ten thousand to keep the costs of perfect cards from being prohibitive. The ultracompact cards would be difficult to repair; faulty cards couldn't be short-circuited because the transit time would be too long; the whole card would have to be changed. In sum, the Josephson technique fell victim to its own virtue — it was too avant-garde. During this period when the new technique was taking its first steps, techniques involving gallium arsenide were not standing still; they also would benefit from modern etching techniques. Since then, micro- and now nanolithography have made great progress. The idea of supercomputers was oriented toward more efficient machine architectures rather than exotic electronics. Of course, we know now that the real jewel of computers is the personal computer.

When a project fails, dissension comes to light, and IBM, in spite of its sophisticated management techniques and control of its personnel, was no exception to the rule. Sadeg Faris, one of the specialists on the superconductor project, someone who had spent his entire career with IBM, decided to put together some electronics based on collections of Josephson junctions, even though the company had given up on them. With the help of military contracts he created a company, Hypres, devoted to the development of Josephson circuits, and he soon announced its first products. They were immediately subject to ferocious criticism by his former colleagues at a conference on applications of superconductivity. Nevertheless, in 1986, after several years of effort, Faris put an oscilloscope on the market, the fastest in the world, with a resolution of 5 picoseconds (1 ps $= 10^{-12}$ s); he sold only twenty of them. The heart of the instrument was a microcircuit containing a number of Josephson junctions made from niobium, which permitted very rapid sampling of a signal. Each fragment of the signal was stored and processed to provide one pixel, one point of the display on the screen. Since niobium has a critical temperature of 13K, the microcircuit did not have to be immersed in liquid helium and was cooled simply by drops of liquid helium. The price of the instrument was very high, $120,000, but its very existence demonstrated that cryogenic electronics could be used outside low-temperature laboratories.

High-current applications of superconductivity require high critical temperatures, but it is important to remark that this is not the principal obstacle to Josephson electronics. After all, the most powerful computers are already cooled by freon to temperatures well below room temperature. The major obstacle to superconducting electronics lies in the fact that we still don't know how to make, in a simple and convincing manner, combinations of Josephson circuits and/or tunnel junctions with a range of applications and ease of use comparable to classic transistors.

The real issue is the nature of the phenomenon, which is very different from the transistor effect. The passage from vacuum tubes to transistor electronics occurred rather easily, because the transistor is like a vacuum triode; both have three wires, one of which determines what goes on with the other two. The passage from semiconductors to superconductors is more delicate, because there is no superconducting arrangement equivalent to a component with three leads. Many have tried, but no one has succeeded, in amplifying a signal with a superconducting component: there is still no "superconducting transistor."

Bringing the superconducting computer project to a halt brought the low-current applications of superconductivity into the limelight. The scope of these applications was much more modest. They are likely to stay in the laboratory and, apart from different kinds of magnetometers, to have little economic importance.

Defining the standard volt using the Josephson effect is a good example. Classically, the standard volt is defined by the behavior of a battery in precisely defined conditions that are not easily reproducible. The AC Josephson effect entails a proportionality between the frequency of an electromagnetic wave incident on the junction and the voltage applied across the two electrodes—here we have the basis of a much more elegant and precise calibration. The proportionality factor is the ratio h/e, Planck's constant divided by the charge of the electron. This is most important for the definition of units because the calibration is thus directly tied to fundamental physical quantities without having to specify the precise conditions under which a battery is operating. While these conditions might be useful to know for technical reasons, they have no conceptual relation whatsoever with the definition of a unit of voltage. Another advantage of the Josephson standard volt: the measurement of a voltage is replaced by the measurement of a frequency. Measurements of time (the inverse of frequency) are among the most precise measurements we can make; the famous atomic clocks regularly reach a precision of better than one part in 10^{14}. The International Bureau of Weights and Measures in Sevres, France, now has a standard volt provided by a network of Josephson junctions built by the National Institute of Standards and Technology in Boulder, Colorado.

In a much more general sense one can say that applications of superconductivity have stayed home in the laboratory. Their utility is hardly measurable in economic terms, and their popularity depends very little on how high the critical temperature is. This remains true even for large-scale applications of superconductivity for particle accelerators or superconducting cavities used in the same machines. The success or failure of these applications has never been crucially dependent on the temperature at which they operate.

The very idea of being able to work at higher temperatures nevertheless continued to motivate research on superconducting materials but with less and less enthusiasm. Even the development of truly original materials whose physical properties alone would justify a program of fundamental research was not sufficient to raise morale.

CHAPTER 19

The Age of Materials

By the end of the 1960s the absence of major results in the race for applications was deepening a post-BCS malaise. The scientists had no real driving goal. A dozen years had gone by since the BCS article had appeared, and perhaps the only project of any sizable scope that was still alive was the superconducting computer. Reaching this goal demanded solving a number of problems, but they weren't compelling; the issues were purely technical. Moreover, since the results of the intermediate steps in this research had not been systematically published, the project contributed nothing to invigorate the brain cells of anyone outside IBM. In other words, there was no longer any obvious direction for development. Superconductivity had become one discipline among many others.

A similar period of stagnation had already occurred between 1911 and 1933, but superconductivity at that point was still completely mysterious, and only a handful of researchers were involved. Now almost everyone thought the essentials of superconductivity were well understood. Although the winds picked up again from time to time along certain secondary research routes, they were not strong enough to drive the hundreds of physicists apparently committed to the field. Some individual careers might blossom, but the buzz among physicists was that "superconductivity was no longer fashionable." Nevertheless, of the laboratories focused on these issues in 1970, not one abandoned the field.

A strongly motivated goal, like the search for high-temperature superconductors, did not necessarily define an epistemologically sound scientific program. High T_c research posed myriad open questions, but it had always been an area in which the most frenzied empiricism reigned. For ten years research on superconductivity could not be justified on its own merits, a fact that all the publications devoted to its glorification pass over in silence. And so, at the start of the 1970s in France, just after university groups and CNRS researchers had made important contributions, this loss of interest was translated into the disappearance of all graduate courses specifically devoted to superconductivity. The fact that the most productive theorists abandoned the field around this time did nothing, however, to reduce the number of articles published. When the goals and social function of science were questioned in the heady days of 1968 in France, some of us could point to this troublesome statistic as a sign of the bureaucratization of science.

What was happening to superconductivity was only one aspect of a more general evolution of condensed-matter physics. As soon as the basic ideas of this field had been clearly posed and understood, what motivation was there to continue to develop, reproduce, and then transmit a well-established understanding? For semiconductors such questions about intellectual goals had always been overshadowed by questions of utility, turning the results of research into widely used, practical devices. Superconductivity, however, could not follow this path, because successful applications had never constituted an appreciable market.

The postwar golden age of superconductivity was over. To borrow Kuhn's terminology, superconductivity had passed imperceptibly from "revolutionary" science to "ordinary" science. The contrast with the study of liquid helium during these same years is particularly clear. Many fewer scientists were working there, and no one claimed to be able to find applications. The research was fruitful, nevertheless, due in large part to the study of the isotope ^3He, in which several superfluid phases had just been discovered at very low temperatures. Superconductivity, on the other hand, was, little by little, first in the United States and then in the rest of the world, becoming a branch of materials research.

Materials research represents the revenge of the chemists after years of unabashed domination by the condensed-matter physicists. It was in this spirit, moreover, that the field was launched in the United States at the end of the 1960s. In France it arrived as usual with some delay. Adapted to the French context, it was a corrective to the narrow conception of science policy that had been dominant, a policy conceived primarily to emphasize well-defined technological careers (e.g., big computers, applications of nuclear physics, space). Materials research tries to produce new materials and to study their physical properties with an eye toward certain kinds of potential applications. The definition is intentionally fuzzy, as it is for categories that are justified by considerations of the politics of science rather than the science itself. The emergence of the theme of "materials" allowed a compromise between the demands of fundamental research and those of technical studies. The clash between fundamental science and applied science was avoided, but the physics of materials research ran the risk of being reduced to characterizing materials synthesized by chemists. Measurements with no guiding principles should hardly be baptized as science.

The listlessness of superconductivity and its drift toward materials research were in large measure the specific fruit of a BCS theory with enormous explanatory power and essentially no predictive power. BCS could not foretell where superconductivity would appear; it provided a global description of superconductivity that did not explain the differences in behavior from one material to another. As the specialists had been saying since 1960: BCS theory doesn't predict anything that is not some manifestation of the energy gap.[1] It's a description of the superconducting state that starts from the hypothesis that there's an attractive interaction between the electrons, but it has no grip on what this interaction is. Thus, it cannot predict the existence of a gap or, a fortiori, the precise value of the gap for a given material. It cannot determine whether or not a material will become superconducting. BCS theory indeed includes an expression with the critical temperature, but only as a parameter; that is, given a value of the critical temperature provided by experiment, it deduces the magnitude of the electron-phonon interaction. In fact, we know how to calculate only very roughly the electron-phonon interaction starting from first principles — the conservation of mass, the principle of least action, the principles of thermodynamics, and so on.[2]

Recognizing this difficulty peculiar to BCS theory, physicists looked for a way to get around the absence of predictive power by posing a slightly different question. Does not BCS implicitly put some upper limit on the temperature at which superconductivity can appear? In other words, is there a temperature above which an ideal material, one with all the properties one could want, could never become a superconductor?

The critical temperature in BCS theory can be expressed relatively simply:

$$T_c = 1.14 \, \Theta_D / e^{(1/N(0)V)}$$

In this formula T_c is the critical temperature, Θ_D is the Debye temperature, and $N(0)V$ is the electron-phonon coupling, where V designates the mean value of the electron-phonon interaction and $N(0)$ is the density of electronic states at the Fermi level. The denominator depends on an exponential, that is, on the constant e, 2.718, raised to the power $1/N(0)V$. A small error in the product $N(0)V$ thus becomes a large uncertainty in the value of the critical temperature. With popular superconductors such as tin, a change of 25 percent in the product $N(0)V$ would triple the value of the critical temperature.

The numerator of this formula indicates that the critical temperature is proportional to the Debye temperature, with a proportionality constant that is almost unity. If this were the entire formula, we could get critical temperatures of hundreds of degrees, because such values are not unusual for Debye temperatures. Unfortunately, we also have to deal with the denominator, the exponential whose argument is likely to be large, because V is small in BCS theory. The denominator is thus likely to be very large, so the division takes us back to very small critical temperatures, the few degrees of ordinary superconductors.

G. M. Eliashberg in the Soviet Union and William L. McMillan in the United States developed models of the interaction between the electrons and the crystal lattice independently of each other. In principle their models can furnish a specific description of all metals and thus an exact calculation of the coupling force, even when it's strong. In particular they take account of the cancellation between the electron-phonon force and the short range Coulomb repulsion between the electrons. (The long-range Coulomb force is shielded by the positive charges of the ions.) Revised in this way, the theory can explain the discrepancies with BCS predictions observed for lead and mercury, discrepancies that are due to the strong coupling between the electrons and the phonons.

The price paid for this improvement was high; it required long numerical calculations specific to each material. This was the era when high-powered computers were first becoming available to physicists outside the bomb centers. McMillan, in collaboration with J. M. Rowell of Bell Laboratories, was able to write a computer program on punch cards to carry out the calculation for lead, a tour de force at the time that today brings forth a little smile. Although the program was designed only for lead, it was used also for other metals by modifying the initial parameters. Each laboratory conserved its packet of cards religiously — one bent card in the deck of an executable version of the program meant instant death for the calculation.

These theoretical considerations left at least one man cold. Following the German empirical tradition, indifferent to theoretical advances, he systematically explored essentially all the elements of the periodic table to see whether they were superconductors. This step finished (after all, there are only ninety-two naturally occurring elements), he pursued, equally systematically, alloys and composites with two elements. This radically changed the dimensions of the search — there are a semi-infinite number of alloys to study. The hero of this research, a chemist among physicists, a physicist among chemists, a man gifted with an unparalleled sense of sarcasm, was Bernd Matthias.

Matthias was born in Frankfurt in 1918. Under the direction of Paul Scherrer, he worked on crystals of barium titanate for his thesis at the Polytechnic Institute of Zurich (ETH), where he received his doctor of science degree in 1943.[3] Barium titanate is a ferroelectric material, with a crystal structure labeled *perovskite*; Matthias devoted much of his life's work to these crystals. Later we will examine perovskites much more closely because, forty years later in the same city of Zurich, perovskites opened the door to high-temperature superconductors. Matthias, however, was not there for the revolution.

Having emigrated to the United States in 1948, Matthias got involved in cryogenics in Chicago in order to study other ferroelectrics and was hired the following year by Bell Laboratories. This is when he began his great work, the systematic exploration of the periodic table. Over the course of thirty years he trained about 150 collaborators. In fact, Matthias led two groups; starting in 1961, he was not only a researcher at Bell on the East Coast but also a professor at the University of California in La Jolla on the West Coast. His double life meant that he was one of the early versions of the peripatetic physicist, always on the road from one coast to the other, his secretary by his side to arrange his routine material and administrative needs. By the time of his sudden death at the end of October 1980, Matthias could be credited with having "discovered" thousands of materials, especially numerous types of ferroelectrics used in piezo-electric devices.

Ferroelectrics are insulators that can become permanently polarized when they are put in an electric field — the positive and negative charges are pulled a small distance apart by the electric field, and they stay that way when the field is removed. The conversion of mechanical energy into electrical energy is called piezo-electricity.

In every physics and chemistry laboratory you will always find attached to a wall somewhere a handy large version of Mendeleyev's periodic table classifying all the elements. The one in Matthias's lab was special. The box corresponding to each element was in fact a little drawer; on its face there were the usual numbers and letters, but inside the box could be found, ready to use, a pure sample of the corresponding element. Any mixture he wanted could be prepared without delay.

Well removed from any kind of theoretical activity, guided only by the rule of absolutely pure empiricism, Matthias, after several years of systematic tries that he claimed were inspired by old German chemical treatises, came upon a totally unexpected result. For the natural elements there was a continuous variation in the temperature at which superconductivity appeared depending on the valence of the element considered. The valence is the number of chemical bonds that an atom must have with other atoms to be in a stable chemical state. Based on his observations, Matthias predicted that alloys with valences around 4.5 or 6.5 would have critical temperatures higher than those of ordinary superconductors.

For a long time Matthias was convinced that a would-be superconductor had to satisfy two conditions: it had to have cubic symmetry, and it had to obey the empirical rule about valences. As he himself admitted many years later, these were two too many conditions. Nevertheless, his empirical rule indicated clearly that the simplifying hypotheses of Bardeen and his colleagues would have to be modified.

The success of the BCS theory had been to show that, whatever the crystal structure and detailed form of the Fermi surface of their electrons, superconducting metals all involved the same interaction. For BCS in its standard version, the Fermi surface was a sphere, and the electron-phonon interaction was isotropic — it was the extreme simplicity of these hypotheses that had induced early skepticism by some condensed-matter physicists. Matthias was interested only in what BCS theory could not predict: the critical temperature. Moreover, the existence of a correlation between valence and critical temperature raised questions about BCS as a unifying theory. In directing his efforts toward exactly what this theory was not designed to do,

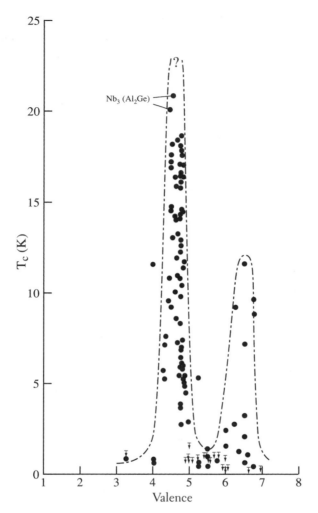

The critical temperature of A15 alloys as a function of the number of valence electrons in the atom. This graph perfectly illustrates the type of empirical research on materials that Bernd Matthias carried on for twenty-five years with remarkable success. These A15 compounds reached critical temperatures up to about 23K, a record level until the new high T_c ceramics. These materials are still used in superconducting windings for magnets. Note that the highest values correspond to valences of about 4.5 and 6.5.

Matthias was ostensibly playing the role of iconoclast, a role that he found not at all displeasing. Indeed, his motivation in the quest for higher-temperature superconductors was to make a significant discovery that owed nothing to BCS.

With this criterion Matthias and his group achieved the goal of his entire enterprise: to be the leader in the race toward high-temperature superconductors. The composite superconductors that have the same crystal structure as tungsten in its

β phase, known by the name A15, provide a particularly compelling example of the validity of his predictions. Matthias discovered around 1964 that a composite of niobium and tin, Nb_3Sn, reached 18.5K. Several years later it was observed that this composite had a critical field of 8.8 teslas — that is, 180,000 times higher than the earth's magnetic field and quite a bit more than the two unhappy teslas of the most sophisticated electromagnets cooled by circulating water. With this composite superconductor he was finally going to make superconducting windings that would create magnetic fields much higher than those of early electromagnets. In fact, in the middle of the 1960s the first superconducting coils appeared in the scientific marketplace. Five teslas were possible, then ten teslas. At the end of the 1960s, at a conference at Stanford University before a thousand specialists from all over the world, Matthias could do his great skeptic number in the tone of the conqueror; his antitheoretical demagogy was always popular with experimentalists. The talk was guaranteed to be a success, because there are so many more experimentalists than theorists, so experimentalists are always in the majority at conferences.

In spite of his grandstanding, Matthias nevertheless had to recognize how much he owed to theory. The theory to which he was most indebted was not BCS, however, but Ginzburg-Landau, which eventually led to the understanding that there were, in fact, two types of superconductors.

With his empirical rules Matthias had managed to increase critical temperatures. As we have seen, however, the most sophisticated applications of BCS theory provided nothing but a framework for considering the question of why superconductivity could not be found at higher temperatures. When Matthias died, the world record for critical temperatures was the one he had helped establish: 23K for a composite of niobium, aluminum, and germanium. It was this result that Bardeen cited on December 11, 1972, when he received his second Nobel Prize, this time with Cooper and Schrieffer, for superconductivity.

Matthias was not the only solid-state physicist-chemist interested in superconductors, but he took care to place his work in a clear context, something his colleagues were not concerned with. And so he could affirm that "previous review papers and compilations of superconductors have generally been descriptive in nature and presented very few conclusions."[4] In the 1963 review article from which this citation has been extracted, he made his concerns about BCS theory quite explicit. He remarked that the history of the isotope effect quite naturally raised some doubts about whether the electron-phonon interaction is the "universal cause of superconductivity": "When the existence and behavior of superconductors is the object of systematic investigation, one realizes that they arrange themselves naturally into three different groups: the elements that are not transition elements, the transition elements and the intermetallic compounds. Conclusions based on one of the three groups should not be generalized to the others."

In the periodic table the transition elements are three columns to the left of the precious metals (the d shell is not complete, in contrast to the alkali metals in the first column; for transition metals the Fermi level is within the d shell). High temperatures are often needed to make these materials, and they are very reactive. The availability of electric arc furnaces has proved decisive.

In fact, once some of the experiments were redone, the isotope effect, the dependence of the critical temperature on the square root of the atomic mass for different isotopes of the same element, had really been confirmed only in mercury, zinc, cadmium, tin, thallium, and lead. The results were in good accord with BCS, but, as Matthias observed, none of these elements is a transition element. They can all be found two columns to the right in the periodic table, just to the right of the precious metals (gold, silver, and copper). What's more, for superconducting transition elements such as ruthenium and osmium, there is no isotope effect. For titanium, another transition metal, the isotope effect is quite variable, but that's due to the presence of impurities.[5] For the intermetallic superconducting compounds such as Nb_3Sn, so dear to Matthias, with a critical temperature that seemed very high by the standards of the 1960s, changing the mass of the tin yields an isotope effect ten times weaker than theory predicts. Citing dozens of compounds, Matthias showed that the simple rule predicted by BCS was only rarely observed, even with modifications that took account of strong electron-phonon coupling.

Was there some other mechanism operating, something beyond the electron-phonon interaction, which was really responsible for superconductivity? The results found by Matthias's groups, as well as the great mass of those obtained by chemists almost everywhere, provided a good demonstration that BCS theory was not appropriate for all superconducting materials. Its simple assumptions about the shape of the Fermi surface were particularly problematic. BCS theory could be extended to exotic materials by making more realistic starting assumptions, but the calculations would then be much longer, and theorists were not convinced they would be useful. The example of McMillan in this field, and of many theorists working on semiconductors, would hardly encourage one to set off in this direction; they had to make innumerable calculations of band structures with adjustable parameters to fit pre-existing experimental data. After all, BCS themselves had clearly indicated that any attractive interaction at all could explain superconductivity, and they were the first to recognize the limitations of their theory.

After a success as spectacular as that of Bardeen and his colleagues, the tendency had been to believe that superconductivity could continue to progress without having to be bothered with the vulgar preoccupations of other condensed-matter experimentalists, who were always fixed on the detailed structures of their samples. What Matthias did that was truly noteworthy was to question the paradigms of a scientific community.

The absence of predictive power in the theory thus gave free rein to the chemists, who, little by little, came to the forefront of superconductivity by proposing a host of new compounds for examination by physicists. This, incidentally, reinforced the tendency to file research on superconductivity under the heading of the physics of materials. The reasons that the chemists synthesized new compounds were often far removed from the interests of physicists involved in superconductivity. The order in which these materials were studied corresponded not at all to any internal logic related to the evolution of superconductivity. It would be equally exaggerated to think that the sole ambition of the physicists in this field was to raise the critical temperature. The harvest of superconducting materials was sufficiently rich that studying them often brought opportunities for refining the theory.

The idea of playing with the Fermi surface naturally comes to mind in connection with the "high critical temperatures at the end of the 1950s." All these materials have the structure of tungsten in its β phase. This means that the atoms of niobium (Nb) in Nb_3Sn (T_c = 17 to 18K) or those of vanadium in V_3Ga (16K) are in a cubic lattice with two atoms per element of the lattice. The atoms of transition elements form orthogonal chains among themselves. Experiments show that, on the one hand, the highest critical temperatures are obtained when these chains are not broken by atoms of another element and, on the other, that the critical temperature is not lowered when the atoms of another element are substituted for those that are not part of the chains. Thus, the compound $Nb_3(Al_{0.8}Ge_{0.2})$ has a critical temperature of 20K. Such results show that the presence of atoms of transition elements along with the anisotropy of the material each play an important role in raising the critical temperature. Jacques Friedel and his students proposed a much discussed model that described this situation.[6]

Again, it was in two transition metals, ruthenium and osmium, that Matthias and his collaborators demonstrated that the isotope effect was almost absent. After an initial attitude of disbelief the theorists explained that the absence resulted from the cancellation of the electron-phonon coupling by the electrostatic repulsion between the electrons. This repulsion is stronger for d shell electrons (those responsible for conduction in transition metals) than for the p shell electrons (the superconducting electrons at low temperature).

Just before he died, Matthias was finally obliged to admit that he had really stayed within a BCS framework. He recognized that, after countless compounds had been studied by hundreds of researchers from all countries, the result of his work was "a substantial improvement in the detailed BCS expression for the critical temperature." He was disappointed that, in spite of everything, his work amounted to a confirmation of the core of the theory. But he could take pleasure in the fact that, although he and his colleagues "did not succeed in finding an exception to the basic electron-phonon mechanism, at least we made the theory more responsive to the experimental facts."[7]

A number of strongly anisotropic materials (materials that are not spherically symmetric) had been synthesized and studied. Physicists involved in superconductivity got to know their colleagues in chemistry who had christened themselves solid-state chemists and seemed capable of designing structures that were more and more anisotropic. "Two-dimensional" superconductors were discovered, in which Cooper pairs formed and moved about preferentially in layers only one or two atoms thick. The dichalcogenides such as $NbSe_2$ and TaS_2 became fashionable. It is really quite spectacular that TaS_2 doesn't mind having a layer of organic molecules interposed between two of its own thin sheets. It's possible to vary the anisotropy of a complete family of compounds by putting layers of organic molecules several tens of angstroms thick between layers that are themselves only a few angstroms thick, like a club sandwich.

Soon materials became available that were effectively one-dimensional, in which electrical conductivity took place along chains of atoms that were more or less in straight lines. From a purely phenomenological point of view, such materials posed no special problem. Von Laue had introduced anisotropy into the London equations in 1948, replacing the simple number representing the penetration depth for anisotropic conductors with a quantity that varied with direction in the sample.[8] The same thing happened to the effective mass in the Ginzburg-Landau equations.

The irony in this entire quest for strongly anisotropic materials is that the young physicists of the 1960s and 1970s were simply following in the footsteps of their forebears when they tried to "explain" superconductivity by linear chains of electrons, as people such as Frenkel and Fröhlich had tried previously. In any case the record for the highest critical temperature remained the one that came out of Matthias's work.

These studies required precise knowledge of the crystalline structure of anisotropic compounds. As long as a spherical Fermi surface sufficed, crystal structure had not seemed important, and the crystallographers had been treated with disdain by the superconductivity community. Now they benefited from a more objective evaluation of their merits and their new tools for generating X rays: the first-generation accelerators used in nuclear and particle physics, including cyclotrons, synchrotrons, and storage rings for accelerated electrons, that were no longer at the cutting edge in these fields. Because these accelerators were circular, intense beams of low energy, highly monochromatic, radiation, mostly X rays, were emitted tangential to the rings. In the competition that opened up to characterize precisely the distortions induced by charge density waves in anisotropic conductors — names of old accelerator labs such as Brookhaven, Stanford, and Orsay[9] — reappeared, this time as synchrotron radiation laboratories. Neutrons from nuclear reactors also became part of the crystallographer's tool kit.

In order to study the structure of materials, probes with wavelengths comparable to the distance between atoms are necessary. According to quantum mechanics, particles have wavelengths inversely proportional to their momentum, and the low-energy neutrons emitted by nuclear reactors have the right wavelengths. Neutrons, which are much more sensitive than X rays to scattering by phonons, can reveal very slight changes in atomic structure. By varying the temperature, it became possible to study structural transitions and to describe them in terms of low-frequency phonon modes, the so-called soft phonons. Most of the time these were second-order transitions. Quantitative description relied on a theory that we have already encountered: Landau's theory of phase transitions. In the case of structural transitions the order parameter no longer represents the density of Cooper pairs, as it does in superconductors. Instead, it is a quantity tied to the amplitude of atomic vibrations or to the atomic displacements associated with this transition.

To evaluate the power of these new techniques and to clarify the interpretation of the results in terms of the theory of phase transitions, the physicists used substances with known properties. The idea was to have samples in which the phase transition could be observed by measuring a macroscopic physical quantity such as the dielectric constant of an insulator and then to compare these measurements with the information on the atomic scale provided by neutron scattering. This is why the crystals that we'll encounter in the next chapter, the perovskites that Matthias was so enamored of, were quite naturally among the first materials studied. Perovskites often have ferroelectric properties that appear at temperatures that are not so low, generally above liquid nitrogen in fact, and thus are relatively straightforward to attain.

The results obtained for strontium titanate ($SrTiO_3$) are a good illustration of what these techniques can provide. Optical and X-ray measurements around 110K reveal that this compound passes from a perovskite phase with a cubic structure at high temperatures to a low-temperature phase with a tetragonal structure. The elastic constants of the material change, even discontinuously, at 110K.[10]

The accompanying explanatory paragraphs describe the geometry of this transformation, which is typical of all these materials. Those readers who have never thought of crystals as the stuff of dreams might want to skip it entirely.

Structural studies of $SrTiO_3$ showed that the phase change at 110K was accompanied by the total disappearance of soft phonons propagating in a definite direction and by the stretching of the sides of the basic cube of the crystalline lattice, the unit cell. The stretching was very modest, perhaps a part in a thousand; only the oxygen atoms were displaced.[11] This was not detectable with X rays, but it could be inferred indirectly,[12] before direct measurements were made with neutrons.[13] The lengthening was accompanied by a rotation of the octahedron centered on the titanium atom,

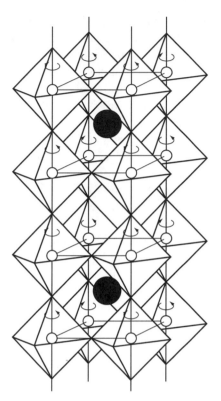

The structure of $S_rT_iO_3$ explained in accompanying text. The strontium atoms are shown as large black circles, the titanium atoms as small white circles. Oxygen atoms (not shown) occur at each point where two octahedrons touch, midway along each side of the cubes.

whose peaks are atoms of oxygen. The rotation is made around the vertical axis in the figure, labeled [100] by the crystallographers. Since some of the oxygen atoms belong to two unit cells, the sense of rotation is inverted in passing from one unit cell to the next, as shown by the arrows in the figure. Because of the rotation, the unit cell cannot be just a slightly extended version of the original unit cell but must be, rather, the overlap of two consecutive unit cells. The unit cell of the crystal lattice is said to be "duplicated" in the transition, and that is why two consecutive cubes are shown in the figure. Finally, we should note that the oxygen atoms, which at temperatures above 110K are aligned in totally straight vertical chains, are no longer so rigorously aligned at lower temperatures but, instead, form zigzag chains like a folding measuring tape that is not completely unfolded.

All these features visible in the phase transitions of perovskites — the disappearance of soft (low-energy) phonons, the doubling of the unit cell, and the deformation of the straight chains of oxygen atoms—will appear in our account once more when we encounter the superconducting oxides discovered in 1986. They too are perovskites. More generally, soft phonons and structural transformations are important in the mechanisms responsible for superconductivity, as they are in many other phase transitions in solids. Soft phonons and structural transformations became observable over the course of the 1970s; the price was the utilization of major equipment, synchrotrons and reactors. In France special reactors were constructed in Grenoble and Saclay. The style of research around these big machines quite naturally followed the high-energy physics model—the industrialization of intellectual work in materials research became more and more pronounced.

For condensed-matter physics the costs of developing and operating these large pieces of equipment were incomparably greater than the costs of the artisanal research carried on in laboratories in which materials were prepared and characterized by other methods. At the same time, research budgets were in a period of stagnation, accentuated by the oil crisis of 1974. Such a large fraction of the research budgets was devoted to heavy equipment that smaller laboratories in several countries, particularly in France, laboratories devoted to fundamental research on the new materials, were left impoverished.

The preparation of samples for the new experiments, an activity less visible than commissioning reactors, often represented a tour de force. It is generally extremely difficult to obtain good monocrystals of anisotropic materials with dimensions large enough to do the measurements. This is especially the case when the chemical bonds in the direction with little conductivity are reduced to electrostatic Van der Waals forces. In order to have a large distance between layers or chains, the binding force has to be weak. Happy is the investigator in possession of a monocrystalline sample of a unidimensional material one millimeter in cross-section. More often these materials are like a tuft of hair, not exactly the ideal sample for application of a magnetic field that must be rigorously parallel to its axis. Orienting such a crystal relative to a beam of X rays or neutrons is no easier.

From three- to two- to one-dimensional materials—why not zero dimensions, structures that looked like metals, but with free electrons that seemed so heavy that they were practically immobile? The first of such materials, $PbMo_6S_8$, appeared in 1971. It was the prototype of a whole series of materials called the phases of Chevrel, after the young French chemist who had just synthesized it.[14] The lead atom in this compound is at the center of a cubic cage with an atom of sulfur at each corner and an atom of molybdenum at the center of each face (thus the eight sulfurs and six molybdenums in the chemical formula). Its critical temperature was 14.4 K, and its critical magnetic field was larger than 60 teslas, both large enough to make it a serious competitor to the superconducting niobium wires used in the production of high magnetic fields. The properties of these phases of Chevrel compounds seemed sufficiently promising that a program for producing such wires was envisaged in France.

Matthias recognized that this material was the first true ternary (three-element) compound, because the ratio of elements in the material was just what the chemical formula prescribed. There was nothing extra whose sole purpose would be to induce deformations in the lattice favorable to the appearance of superconductivity. At the same time that anisotropic structures were attracting many investigators, the number of chemical compounds that were candidates for superconductivity was becoming potentially infinite. If ternary compounds were possible, why not quaternary? Perhaps, someday. The day arrived with superconducting oxides.

During this entire period teams of French solid-state chemists were responsible for many discoveries of superconducting compounds with low dimensionality. The laboratories involved included Talence, near Bordeaux (Paul Hagenmuller), Rennes (Marcel Sergent), and Nantes (Jean Rouxel). At least at the beginning the chemists were interested in superconductivity only as a by-product, but they needed collaborators from physics to decide whether or not their new compounds were superconductors. The physicists, however, tended to regard them and their recipes with disdain, so collaborations between French chemists and physicists were infrequent. The consequence was that most of the physics results were obtained outside France. For example, the superconductivity of the Chevrel phases was discovered not in France but in the United States, inevitably by the consortium of Bell Labs and Matthias's La Jolla lab.[15] Another example was the discovery, in Geneva, that these compounds remain superconducting even in strong magnetic fields, several tens of teslas.[16]

The continuous stream of important contributions from the French chemists owes much to the personality of Paul Hagenmuller. He revived this discipline, which had suffered greatly during the war, with a strong dose of fresh air from foreign countries. Such behavior was hardly typical for someone set up not in Paris but in Bordeaux. In order to fund his laboratory he searched energetically to find industrial applications, civil (and military), for the work of his group. Even at the beginning of the 1960s, and even for chemists, who have a longer history of dealing with industry than physicists (by roughly one world war), his activities were unusual. Yet he didn't sacrifice the quality of his output to the demands of applications, something even less typical. Hagenmuller knew how to keep his best students, and he knew how to attract vigorous researchers to his laboratory.

One-dimensional materials have a special place even among superconductors with low dimensionality which Hagenmuller and his colleagues were concentrating on. Two factors strongly motivated researchers in this area. On the one hand, they hoped that they might somehow find a material in which the electron-phonon interaction was particularly strong. On the other hand, they all secretly hoped they would come upon a superconductor in which phonons were unimportant, because, theoretically, all that was needed was some kind of effective attraction between two electrons. Theory did not require that the attraction arise from phonons, and the structure of the pair itself need not come from two electrons moving in a coordinated way in space. Perhaps the spins of the electrons would provide the pairing force.

When a pair of electrons with their spins aligned parallel to each other is placed in a magnetic field that is not too strong, quantum mechanics suggests that it will have three different energy states, corresponding to the three possible orientations the total spin of the pair is allowed to have relative to the magnetic field direction. These pairs are thus labeled triplets, in contrast to Cooper pairs with the two spins aligned opposite to each other. The total spin of a Cooper pair is zero, and only one energy state is possible in a magnetic field. Maybe they wouldn't hit the jackpot of high critical temperatures, but the one-dimensional materials seemed at least to be the kind of material that might yield triplet pairs.

Organic materials often consist of long chains of molecules, with electrons that move about freely over their entire length. As early as 1964, W. A. Little, and then V. L. Ginzburg were speculating about the possibility of organic superconductors and publishing frequently.[17] In thinking about the nature of superconductors this way, they were following in the footsteps of Fritz London, and that's the way that Little saw things as well. William Fairbank, a bold physicist and a specialist in low temperatures, encouraged Little, his colleague at Stanford, to consider the evolution of London's thought, from covalent bonds and superconductivity to large organic molecules and the diamagnetism of benzene.[18]

The benzene molecule is a regular closed hexagon. In looking at a kind of X ray of a DNA molecule that had the form of a long loop folded back on itself, Little made a connection with benzene and superconductivity. Electrons traveling together in organic molecules might be superconducting, and a very rough calculation suggested to him that the critical temperature might be 2000K!

Little and Ginzburg's many publications popularized the idea of looking for superconducting macromolecules, but physicists generally greeted the idea with a knowing smile. It was worse at the 1969 conference on superconductivity at Stanford. Just before the conference, the periodical *Soviet Union*, published in Moscow but distributed in the United States to show off the merits of the USSR, published an interview with Ginzburg about organic superconductivity. Ginzburg ventured a prediction that an organic material might have a critical temperature as high as 40K. What appeared in print, if we remember correctly, was 400K. Whatever, an extra zero or an error in units, the entire conference broke up laughing over this unscheduled communication.

In spite of this, the conference on organic superconductivity did take place. Matthias said it was the first conference ever on materials that didn't exist. After the Nth review article on the quest for high-temperature superconductors, Matthias tightened the screws a bit: "I don't know of any other field of physics in which there have been so many predictions without a single experimental result. . . . The deluge of pointless speculations that overwhelms us today on every side can only increase the credibility gap instead of the energy gap."[19] His conclusion was a bit grandiose:

In the right hands, BCS theory provides a splendid approach to the description and explanation of superconductivity. However, its development has gradually acquired many characteristics that bring to mind Goethe's sorcerer's apprentice:

> Herr, die Not ist gross
> Die ich rief, die Geister
> Werd ich nun nicht los.

> [Master, great is my distress today
> The phantoms that I have summoned
> Just will not go away.]

Little took his time and, in the April 1972 issue of *Physics Today*, he replied: "The article of Bernd Matthias on 'The search for high temperature superconductors' . . . unfortunately gives a rather parochial view of the field. The latter part of the article, which attempts to deal with the most challenging questions of higher temperature superconductivity, is particularly disappointing. This results from a number of imprecise statements, omissions, and trivial errors, some of which we wish to point out here." Little went on the attack. Far from judging organic superconductivity as utopian, he emphasized that numerous ferroelectric phase transitions in organic materials were an inducement to continue along this path.[20] In response to Matthias, who for once had taken cover behind the theoreticians to support his point of view, Little observed that McMillan himself had clearly indicated the limits of his work. The limit of 40K for type A15 compounds that he fixed in his revised formalism for the electron-phonon interaction should not be interpreted as a physical barrier to higher-temperature superconductivity. Rather, it was a limitation of a formalism that assumed that the phonon frequency could be infinitesimally small. Yet, according to McMillan: "Of course, this is not the case. We are likely to drive some phonon mode unstable, so that the metal prefers a different crystal lattice, before the average phonon frequency is decreased very far."[21]

As we've seen, this is just what happens in prerovskites such as $SrTiO_3$. To finish up, Little announced the publication of the proceedings of the famous conference on organic superconductors that Matthias had scorned.

A few months later an American, Alan J. Heeger, a physics entreprenueur and an associate of the chemist Anthony J. Garito, thought he had discovered the first member of a whole new class of materials. He showed what he believed was evidence for extremely high conductivity in a sample of organic material at 60K, a conductivity five hundred times larger than its conductivity at room temperature.[22] The chemical formula of the sample was TTF-TCNQ, tetrathiofulvalene-tetracyanoquinodimethane. This formula represents linear organic chains.[23] His result, which surpassed the wildest dreams of more ordinary physicists, was a sensation at the American Physical Society meeting where it was announced. He had not found superconductivity, but it seemed to be right around the corner. His experiment, however, was not reproducible: of seventy samples only three seemed to have this remarkably high conductivity in a narrow range of temperature. Yet the experimenters, remembering the remarks of Peierls about the structural instability of one-dimensional conductors, decided that the system was just on the edge of superconductivity. The crystal had been distorted, and the unit cell had been duplicated. Bardeen himself was enthusiastic about it, and he recalled Fröhlich's suggestion. In fact, the measurement was grossly wrong.

It seemed to be a totally ordinary experiment, a measurement with four wires: two of them brought current to the sample, and the voltage was measured across the two others. If the resistance was very small — that is, the conductivity high — the measured voltage would be small even with a high current ($V = IR$). The chemists found a small voltage for TTFNTCNQ, but the measurement was wrong because they had not taken care to ensure that the electrical contacts were properly made where the current actually existed. If no current exists in a certain region, the voltage will be zero regardless of the resistance. This is just what had happened in this sample: the current existed in one part of the long chain, but the voltage was measured in a different chain where there was no current.

Once the mistake was corrected, the enormous increase in conductivity was reduced to a slight rise; it did, indeed, signal a rearrangement of the atoms, but it had nothing to do with superconductivity. In the meantime the National Science Foundation had awarded $500,000 to the laboratory where the measurements were made.

Another quite surprising avenue of investigation was also explored: nonorganic polymers, compounds with neither hydrogen nor carbon. An American group found that a polymer made of sulfur and nitrogen $(SN)_x$, where x could be arbitrarily large, was superconducting at a few kelvins.[24] The hope of finding organic superconducting molecules was not totally unfounded. It was in Europe that organic superconductivity would finally be manifested.

Exerting high pressure on a material at low temperatures is one way to provoke a change in structure. Denis Jérôme at Orsay led a team of experimenters studying anisotropic materials this way for a number of years. They showed, after a number of setbacks and unfounded celebrations, that a particular kind of organic salt $(TMTSF)_2$ synthesized by a Dane, K. Bechgard, did in fact become superconducting.[25] The critical temperature of Bechgard's salts was not particularly high, however, and, more generally, the number of superconducting organic materials has remained small up to the present time.

Simultaneously, another class of materials, chemically quite different, was discovered by a German group. Some examples are UBe_{13},[26] $CeCu_2Si_2$,[27] UPt_3,[28]

U_6Fe,[29] and YBe_{13}, all superconducting. It was not the critical temperature of these materials — less than 1K — that attracted attention. What was interesting is that they all have critical magnetic fields that are quite high and, in addition, specific heats that are extraordinarily high, a hundred to a thousand times higher than ordinary metals. At low temperatures the specific heat is due essentially to the electrons and not to the phonons. The large values for the specific heat thus means that it takes a lot of added energy to make the electrons move around faster, because their effective mass becomes very large, a few hundred times normal. This is because the conduction electrons in these compounds are closely tied to the (much heavier) ions, and they acquired the label "heavy fermions." In technical language one says that they belong to a very narrow conduction band with a density that peaks very close to the Fermi surface. This characteristic led the theorists to think about superconductivity. Once again they proposed that a pairing mechanism that did not involve an electron-phonon interaction is at work here.

The calm routine of superconductivity research looked as if it wasn't going to change, when a rumor suddenly began to circulate at the end of the fall of 1986: a Swiss at IBM in Zurich had found superconducting oxides at several tens of kelvins. Another hoax? Everyone seemed to have fallen for this kind of trap before: the Germans a long time ago, the Americans, the French, and the Russians. The Swiss had not; they weren't the type. But the rumor was insistent; it was serious. They did indeed seem to have attained some tens of kelvins for the critical temperature, and it was said also that the oxide was easy to make. All at once all the disparate ideas that had been bandied about over the past fifteen years were going to be measured with the new meter stick of these unexpected materials. Listening to this buzz in the corridors, could one have guessed that several months later superconductivity was going to make the front page of the most famous of American newsweeklies?

CHAPTER 20

A Swiss Revolution
The Superconducting Oxides

Ever since superconductivity became so intertwined in the hodgepodge of materials research, it had become difficult for a researcher to decide how to proceed. The search for high-temperature superconductors had no program. In the literal sense there was no scientific policy for superconductors, even though at the same time it was possible to define a policy for applications of superconductivity starting from the clearly established basis of BCS theory and the Josephson effects. What we're about to talk about now should thus be considered as a miniscule part of the research activity on superconducting materials, one path among many with no particular connection to other paths and no expectation of leading to some kind of breakthrough. This particular trail turned out to be the right one, although certain steps seemed quite risky.

In the mid-1980s the record for the highest-temperature superconductor had been held since 1973 by three researchers from Bell Labs. They had shown that a thin film of Nb_3Ge obtained by evaporating niobium and germanium together became superconducting at 23.2K.[1] Once more it was a compound with an A15 structure. The increase of 0.9K over the preceding record with the same compound had been greeted with polite but measured interest. It had become evident to one and all that the A15 track was not about to yield much higher temperatures. All efforts since that time had been guided by imprecise arguments, reasoning that was so general that nothing practical could be deduced from it. The chemists had taken their revenge, not only experimentally by synthesizing samples but also by taking over, little by little, the classic techniques of physical measurements — the instruments were now available off the shelf. Physicists were losing control of their tiny empire because it was the chemists, with their empirical reasoning and arguments from analogy, who were making the news.

The condensed-matter physicists had scarcely any idea of how to advance beyond the model of free electrons in an isotropic crystal that had been proposed by Bloch in 1928. To be clear, even today this model is a perfectly good first approximation for most transport properties in solids and for the theory of superconductivity itself. In Bloch's model an electron can range over the entire crystal; it is described by a wave function, called a *Bloch function*, which is the product of a plane wave and another function with the periodicity of the crystal. Starting in the 1970s, this representation was no longer sufficient to nourish new ideas at the cutting edge of the field. It was so general that it was fundamentally incapable of describing the specific behavior of one of those numerous solids that went through several phase transitions as they were cooled. The structural transformations, then, are often correlated with considerable anisotropy, which is poorly described by models of more or less free electrons derived from the Bloch model. A description in terms of chemical bonds, however, takes this anisotropy into account quite naturally.

It is ironic to note that, with the crucial exception of the discovery of electron pair condensation, the early days of superconductivity seemed to have returned for an encore. This was the era when, in Leiden, Einstein, Ehrenfest, and Kamerlingh Onnes gave free rein to their imagination to invent collisionless electron trajectories. A half-century later, with all the results of quantum mechanics as support, it was a search for the right electron orbitals, the ones that would lead to a system of chemical bonds for which cooling would produce a favorable structural transformation. Few physicists were really prepared to launch such an enterprise. They had accepted the language of the chemists only when the superconductivity of organic macromolecules was the issue.

In these molecular models that the chemists talk about, one distinguishes between the electrons that stay attached to atoms and those that can move around. Chemists and physicists alike call an electron that can move along a chain of atoms a π electron. At this level of generality the expectation that two disciplines would make use of the same concept is almost self-evident. The designation π *electrons* suggests different images and theoretical constructs, however, for the two communities. Agreement on the word signifies only that both want to talk about the same thing, and that's already not a bad start. Nevertheless, physicists continue to represent conduction phenomena in velocity space (or momentum space, which is almost the same thing).

The almost involuntary joint effort of chemists and physicists is at the origin of the quest for new superconductors. The year 1973, when, for the last time, an A15 compound raised the record value of the critical temperature, is also the year in which superconducting oxides entered the game. At La Jolla (the University of California at San Diego) David Johnston, a student of W. H. Zachariasen, a colleague of Matthias, observed that the compound $LiTi_2O_4$ becomes a superconductor at 13.7K.[2] As he says in the conclusion of his article in *Materials Research Bulletin*: "$Li_{1+x}Ti_{2-x}O_4$ is the first superconductor at high T_c ([*sic*] for a reader of the 2000s) that contains oxygen as a primary constituent. Previously, the highest T_c reported for an oxygen compound was 6–7K for $Rb_{0.2}WO_3$." The concept of a "high" critical temperature is, like many others, historically relative. Buried in a crazy-quilt publication on the most diverse materials, the article was ignored, even though Matthias had recommended its publication.

With hindsight we can see that the next step toward the discovery of superconducting oxides, with critical temperatures that were truly high, was taken by Arthur W. Sleight and his team two years later at the Dupont laboratory in Wilmington, Delaware. This time it was not one but a series of oxides that were superconductors. They follow the general formula $BaPb_{(1-x)}Bi_xO_3$, where the four elements are barium, lead, bismuth, and oxygen. The x can vary between zero and one, which means that these compounds can be made with a chemical composition that can be varied continuously from $BaPbO_3$ to $BaBiO_3$ — that is, substituting bismuth for lead. Between these two extremes the critical temperature reaches a maximum of 13K.[3] That's 12K below the record for the time, but it represents a temperature three to five times larger than expected from the McMillan formula for a conventional superconductor having such a low electron density. That means that the electron-phonon coupling must be extremely strong, since it is the product of the electron density and the coupling strength that fixes the critical temperature.

This wasn't enough to capture the attention of the theorists. But these oxides had another remarkable characteristic. The chemical composition of the compound with the highest critical temperature, $BaPb_{(1-x)}Bi_xO_3$ with x = 0.3, is similar to one that has a metal-insulator transition. Finally, although no one attached any special significance to it, it had a perovskite crystal structure. This meant little to the typical physicist in 1975, even though the widely read *Physics Today* had devoted an article to perovskites two years previously.[4]

These indicators of coming events happened in the United States, but they were blithely ignored for almost ten years. We shouldn't be too surprised, then, that the discovery of high-temperature superconductivity was the fruit of a marginal enterprise started in Switzerland by an IBM physicist close to retirement, completely divorced from the pressure of scientific competition. A far cry also from the vaunted precocity of genius.

It was common practice at IBM to give a few particularly outstanding researchers total liberty in their choice of research; they could join the elite club of IBM "fellows," with only a dozen members. One well-known IBM fellow was Benoit Mandelbrodt, an early promoter of fractals. When Karl Alex Müller turned sixty, he too was given carte blanche for his few remaining active years; he had been director of the IBM laboratory in Zurich-Ruschlikon until then. He had had an honorable career as a physicist, working particularly on the superconductivity of the intermetallic compounds A15. One can only suspect, however, that he became an IBM fellow primarily because Heinrich Rohrer and Gerd Binnig, two members of his laboratory, had won the Nobel Prize for their invention of the scanning tunneling microscope, an honor much rarer at IBM than at Bell Laboratories.

A scanning tunneling microscope produces images of surfaces with a resolution on the scale of atomic dimensions. Its essential component is a tip about one hundred angstroms in diameter placed ten or twenty angstroms from the surface of a sample. A small voltage applied between the tip and the surface produces a current between them via the tunneling effect described previously; the larger the distance, the smaller the current. As the tip scans the irregular surface, the distance to the surface is maintained constant by a piezo-electric quartz crystal, which expands when a high voltage is applied across two faces. The magnitude of the tunneling current determines the voltage necessary to keep the distance constant. The changing values of the tunneling current yield a map of the surface, atom for atom, over the whole region scanned.

Given his new freedom from scientific competition and reporting on results, Müller decided to spend his time on the discouraging if not hopeless problem of high-temperature superconductivity. It was a challenge to himself worthy of Don Quixote. Afterward he spoke of it in the detached manner of the scientist: "When I became an IBM fellow, I had time to do research again. I resumed my work on superconductors again, but decided to move away from A15 intermetallics and search for high T_c s in metallic oxides."[5] Müller's approach was thus radically different from the one so eloquently popularized by Matthias. Rather than launch a gigantic program of systematic empirical exploration, by choosing the metallic oxides as his only domain of exploration he put himself from the get-go into the thick of hypotheses nourished almost entirely by one theorist, hypotheses that are even today somewhat controversial.

His principal source of inspiration was the work of Benoy K. Chakraverty, a theorist in Grenoble, France. Throughout the 1970s Grenoble had remained a vigorous center of research on superconducting materials. In two articles published two years apart Chakraverty considered systems with strongly attractive interactions between the electrons and the phonons.[6] He showed that there could be phase transitions from the superconducting state, not to the usual (normal) conducting state but, rather, to a particular insulating state called *bipolaronic*.

In the middle of the 1970s Chakraverty had been part of a group studying titanium oxides, in which phase transitions between the metallic state and the insulating state occurred as the temperature changed.[7] This work had led them to demonstrate the existence of "bipolarons." In Chakraverty's words: "We can consider entities like localized Cooper pairs (a pair of electrons on neighboring atomic sites, one with spin up and the other with spin down); they are stable as long as the mutual attraction between them due to the deformation they induce exceeds the Coulomb repulsion. . . . These bipolarons are BCS pairs made up of a localized electron and a free electron."[8] Well before this, as we know, McMillan had derived his formula that predicted the value of the critical temperature as a function of the strength of the coupling. This was the basis for the conviction of most physicists (except Matthias, of course) that electron-phonon coupling would never yield a high-temperature superconductor.

Everyone had the same outline in his head. If the coupling is too weak, no pairs will form, and there will be no superconductivity. If the coupling is strong enough, superconductivity will appear, but the critical temperature will not increase if the coupling strength is increased. Along the axis where the coupling strength is increasing, all things being equal otherwise, there would be only one phase transition, the one we have talked about since the beginning of this book, between the normal metallic state and the superconducting state. P. B. Allen and R. Dynes, however, had remarked that this expectation was only an artifact of McMillan's calculation;[9] they expected that T_c would continue to increase with the coupling strength.

Chakraverty emphasized that, on the contrary, as the coupling between electrons and phonons became stronger and stronger, the electrons of the Cooper pairs would localize and form bipolarons. In other words, in a strong coupling regime there would be no free electrons; the superconductor would become an insulator. The physicists' little outline had to be changed: along the axis where the coupling strength was increasing, the new scheme would be the normal metallic state first, then the superconducting state, and then the insulating state. This transition from superconductor to insulator had never before been envisaged.

At about the same time, in a completely different category of materials—compounds rich in boron—the existence of *bipolarons* was well established. These compounds consist of clusters of twelve atoms at the peaks of an icosahedron (a regular polyhedron with twenty sides), where each face is an equilateral triangle. Although each boron atom has five closest neighbors, only three electrons are available for chemical binding, and that is why the three atoms of each face share the other two electrons with each other. The most likely spot to find these two is at the center of the triangle: they form the bipolaron. The two conditions for formation of a bipolaron are both nicely satisfied in these compounds: the Coulomb repulsion is weak in this geometry because all excess charge must be spread over the twelve atoms; in addition, since boron atoms are small, the bipolarons are situated at the center of the faces and strongly coupled to the icosahedron.[10]

According to Müller, Chakraverty's speculations furnished the starting point for a well-lighted path toward higher critical temperatures. In fact, the existence of the highly localized bipolarons in a solid implies also the existence of highly localized negative charges. Each bipolaron thus maintains a permanent electrostatic interaction with the crystalline network that surrounds it. To minimize the energy of this interaction, the ionic network responds to the presence of a bipolaron by deforming a bit, just as a rich chocolate mousse sags slightly under the weight of a small spoon. The bipolaronic insulator is thus the site of local deformations of the ionic network. Now we can see how Chakraverty's work links up with the innumerable searches for mechanisms of high-temperature superconductivity. He proposed a coupling between electrons associated with a structural deformation, and he cleverly showed that it was possible to pass from the superconducting phase to an insulating phase with bipolarons.

It is important to remark here that Chakraverty's paper was one among many and that his speculations required a serious determination on the part of the reader to search out a guiding principle. Chakraverty himself, moreover, had to be especially persistent to get his paper published. His first letter on the subject was published without difficulty in 1979. Nevertheless, as he pointed out at the end of the letter: "Dr. Julius Raninger, a member of my own laboratory, has brought up as many objections as possible to this article, but he has graciously accepted my phase diagram."[11] Chakraverty met much more serious resistance when he tried to publish the calculations that supported his point of view. His article, submitted to *Journal de Physique*, was thrice refused by the anonymous referees before the fourth version was finally accepted, twenty-one months after the first submission.[12] His persistence was not in vain.

One physicist at least, Alex Müller, who had nothing more to prove, found in this article the critical blueprint for his future experiments. He did have a problem, however—who would make his samples? In his Zurich laboratory there was a German chemist, Johannes Georg Bednorz, who didn't get along very well with Rohrer and Binnig, the inventors of the scanning tunneling microscope, and had stopped working with them. He was perfectly happy to join up with Müller, but, at the age of thirty-seven, he wasn't an IBM fellow; he had to account for his activities to the laboratory administration. His collaboration with Müller had little chance of yielding anything but routine results. Müller, a past director, and Bednorz soon agreed, however, to keep the administration completely in the dark about their collaboration. If the project failed, Bednorz could not be blamed. So, Bednorz secretly started to work with Müller in his off-hours.

The two conspirators set about looking for oxides that were potential candidates for strong electron-phonon interactions. The results of the 1973 investigations provided their starting point. They perfected a method for preparing perovskites, because it seemed that in all probability they would have to make many of them, since most metallic oxides had this crystalline structure.[13]

Like Molière's M. Jourdain, who was surprised to hear that he had been speaking prose all his life, we use perovskites every day without realizing it. They're the most abundant minerals on earth, making up more than half the volume of the planet.[14] Geologists have found them interesting for years, but it's hard to find the word in a dictionary. The perovskites are ceramics, compounds of several metallic atoms with a nonmetallic atom, often oxygen. Indeed, this was the case with the mineral that gave them their name. The original perovskite is a calcium titanate, $CaTiO_3$; it was described in 1830 by Gustave Rose, a geologist who named it in honor of his colleague Count Lev Aleksevich von Perovski.[15] The general formula for perovskites, natural or artificial, is ABX_3; clearly, $CaTiO_3$ is an example. The formula describes several hundred materials with all sorts of electrical behaviors; this is what makes perovskites the materials of choice for many electronic components. The original perovskite is an insulator, others are semiconductors or ionic conductors (in which both the electrons and the ions are mobile), and some conduct electricity as well as metals do. Slight modifications to their basic structure can radically change their electrical properties.

The basic structure is very simple. Take a cube, put a large atom of the metallic element A at its center, at each of the eight corners of the cube put an atom of the metallic element B. A and B are cations, that is, positively charged ions. To complete the structure, the positive charge must be neutralized. The recipe: along each edge of the cube, put a negatively charged atom of the element X. If X is oxygen, like Jules Vernes's famous Doctor Ox, you have a perovskite oxide.

As often happens in trying a new recipe, once it's finished you may think you've been taken. You wanted to make the structure of the compound ABX_3, yet you've used twenty-one atoms, and that must mean you have AB_8X_{12}. No! Each edge of the cube is common to four cubes, and each atom B is shared among eight cubes because it finds itself at one of the peaks of the elementary cube. So, twelve edges divided by four gives three X atoms per cube, and eight B atoms divided by eight cubes gives one B per cube. ABX_3 is the right formula for your compound. One concern relieved, but another, more serious, springs up. In following the recipe, indeed you have reduced the field created by the positive charges, but there is little chance that the cancellation is perfect because the electric field created by a charge varies with distance. Don't be upset; this is precisely the charm of the perovskites. The system will spontaneously deform itself until the effective center of the positive charges coincides with that of the negative charges. One more detail: this deformation depends also on the temperature at which you have cooked up your stew.

This ability to change their properties dramatically under deformation is the basis of the wide application of perovskites. The most commonly used is a synthetic, barium titanate, $BaTiO_3$. Its structure is just as we have described in our recipe: the distances between the atoms are such that the effective centers of the positive and negative charges are the same without any deformation. That is not to say that it can't be deformed. Put it in an electric field, and the barium atom will inch away from the center. The positive and negative charges are no longer at the same place, and the crystal is said to be polarized. Take the electric field away, and the crystal will slowly return to its initial perfectly symmetric structure, releasing the mechanical energy stored in deforming itself. This property is used in making capacitors that dampen the occasional high-voltage surges that can destroy vital electronics in televisions and computers. When the voltage pulse arrives, the capacitor deforms and releases the excess charge only very slowly. This is an example of piezo-electricity, which is also used to make microphones and loudspeakers of the same material. Don't think of these perovskite ceramics as a new kind of rubber, however; they are just as hard and breakable as the ceramics you're used to. It takes only a microscopic displacement to neutralize the charges.

Researchers in the laboratory of crystallography and materials science at the University of Caen had an idea for a particularly interesting innovation, with important implications for our story. This group, including C. Michel and B. Raveau, thought about constructing a device for measuring oxygen pressures with perovskites as sensors. Perovskites, like all real crystals, are not perfect; they have faults of various kinds. One of them involves missing oxygen atoms and thus deformations and changes in their electrical properties. Depending on the pressure of ambient oxygen surrounding the perovskite, some of the faults would disappear. The idea of the device was thus to put a perovskite into an atmosphere containing oxygen in such a way that a simple measurement of electrical resistance would indicate the fraction of missing oxygen and thus the pressure of the surrounding oxygen.

Raveau started thinking about a compound well known to chemists, bronze tungsten oxide, Na_xWO_3, in which the binding between the oxygen atoms and the tungsten atoms is such that the tungsten has an effective valence of something between five and six. The distance between two neighboring tungsten atoms is too large to maintain metallic conduction, and so conduction occurs via the intermediary of the electronic orbitals of an oxygen atom. It is easy to imagine that in these conditions the oxygen concentration might well affect the electrical conductivity. Now, shouldn't it be possible to replace the tungsten with copper and do even better? Copper has two oxides, CuO and Cu_2O, thus two oxidation states, both of which are poor conductors. Something like this had been tried before. In Hagenmuller's laboratory in Bordeaux, in 1978, $LaCuO_3$ had been synthesized but only under high oxygen pressure, 65 kilobars, hardly encouraging for the applications the Caen team had in mind.[16] Nevertheless, Raveau and his colleagues set to work, hoping that the synthesis would be easier if some of the copper ions were replaced with alkaline-earth atoms (calcium, strontium, and barium) so that the eventual compound would contain only divalent copper.

They had mixed results. Indeed, it was easier to make compounds by adding the alkaline-earths: it sufficed to mix powders of different oxides in the proportions required by the chemical formula and to heat it in air to about 1000°C. The copper ions, however, still had a mixed valence. The article by Raveau and Michel appeared in the 1984 volume of the Bulletin de chimie minérale, on the compound $La_3Ba_3Cu_6O_{14}$, which was in fact rich in oxygen vacancies.[17] They had measured the resistivity down to 90K. They couldn't go lower because the refrigeration was limited to liquid nitrogen temperature (77K), and the sample was cooled by direct contact. Now, the University of Caen certainly had laboratories that used liquid helium as well as researchers interested in superconductivity. As often happens, however, the various groups were not on good terms with one another, so the resistivity measurements were not done at liquid helium temperatures (4.2K). Raveau says also that he made some attempts in this direction at a laboratory in Orsay, with no result.[18]

Bednorz and Müller had been collaborating already for about two years when, at the end of 1985, they read the article by Michel and Raveau. The chemical formula provided there was a little more complicated than the one for standard perovskite, but, given that the crystal structure seemed similar to those envisaged by Chakraverty, they decided it was worth the trouble to look at the electrical properties of these compounds at low temperatures. In the first phase Bednorz was the important player, since he had to redo the Caen synthesis. Chemistry, we all know, has nothing to do with cooking. Cooks don't like to follow recipes. Surprise — neither do chemists. A chemist much prefers to continue using techniques that are familiar (and the little twists that are part of the game) rather than follow step by step a preparation outline given in an article, however detailed it might be. Let us recognize in passing that physicists have precisely the same attitude. Simply following someone else's instructions rarely leads to understanding, but when it's a question of preparing a sample physicists are often content to do just that "since it's only chemistry."

In any case Bednorz went to work using a method he had really mastered: coprecipitation. Starting from a mixture of aqueous solutions of barium, lanthanum, and copper nitrates in the right proportions, it consists of inducing the formation of the corresponding oxalates by adding oxalic acid. Readers suffering from kidney stones and those afraid to touch food rich in oxalates (e.g., asparagus, sorrel, bananas) can guess what happens next: the oxalates precipitate out in the form of little grains of sand inside a sort of gel. The precipitate is collected and brought to high temperature in a reducing atmosphere. This is when the chemical reactions among the compounds take place and give birth to the desired compound — or else to a different one that closely resembles it but has not at all the same electrical properties. There's a reason why the preparation of ceramics has for so long remained surrounded in mystery. As expected, the preparation of these compounds required severe attention to detail and, in spite of the care taken to control the experimental conditions, most often resulted in a mixture of several phases. All this could easily have ended in a dead end, and in fact the year 1985 ended with no significant outcome. The Christmas break, however, included the last carefree moments the team would have for some time.

Bednorz and Müller saw their efforts bear fruit on January 27, 1986. The automatic system for measuring resistance as a function of temperature, which they had learned how to use, finally delivered a curve that looked like what they expected. The electrical resistance of one sample went up slightly as the temperature decreased only to, all of a sudden, around 30K, start to decrease and get to zero at 10K. At least part of the sample seemed to be superconducting around 30K, if they were interpreting things rightly; then the record for the critical temperature would just have leaped upward by 7K.

This sample, like all the others, contained barium, lanthanum, copper, and oxygen. A new chapter in superconductivity had, it seemed, just begun. Bednorz wanted to pick up the telephone right away and announce the result to the newspapers. Müller, however, was well aware of the disappointment of others who had thought they were seeing superconductivity with their own eyes, only to discover later that nothing of the sort was happening. He convinced his young colleague to repeat the sample preparation to be sure that the effect was reproducible. The Swiss revolution was going to be peaceful. After the first few exciting moments, in fact, they realized that they had several phases of the same compound, possibly even several different compounds, and that what they had did not correspond to what Michel and Raveau had published. In short, Bednorz and Müller had found the right recipe for making their compound, but, if they wanted to keep their reputation, they had to try to clarify the conditions for obtaining superconductivity at 30K.

A plethora of compounds and phases of BaLaCuO (some pronounce it as one word, *balacuoh*) had to be explored, changing the relative proportions of the different elements and determining their crystal structure. It took them three months to convince themselves that they did indeed have a superconductor at 30K and to discover how it was different from what Raveau and Michel had found. During this entire period our two conspirators acted as if nothing unusual were going on, and even the lab directors were not au courant. Then it was time to write up their work. Müller maintained that they should keep a low profile to avoid being scooped while they proceeded with their investigations. Finally, they brought the new lab director, Eric Courtens, into the game. The solution was clear: publish in the *Zeitschrift für Physik*. Its director, W. Buckel, head of the superconductivity group at the Julich laboratory in Germany, would immediately understand the importance of the article and might be asked to be the referee, to eliminate prepublication leaks. Courtens was a member of the editorial board. He participated in the final discussions about the paper with the two experimenters, and he reread the same version that Buckel received on April 17, 1986. It seems that Buckel was forewarned; the paper was, in fact, accepted for publication that same day, although no acceptance date is mentioned in the published version.

The secret was kept. When that issue of *Zeitschrift* arrived in laboratories in September 1986, no one knew that it contained an article that would open new perspectives on the search for high-temperature superconductors. Its title was not exactly eye-catching: "High Temperature Superconductivity Possible in the System BaLaCuO."[19] The article begins, not without irony, with a quote from American physicists: "The empirical search for new materials is at the forefront of supercon-ductivity research."[20] The rest of the introduction contradicts this initial quote and brings out the main ideas that led the authors to their discovery. Several examples are cited to emphasize their hypothesis that, if perovskite oxides can be supercon-ducting, it's because of the strong electron-phonon coupling. They also explain in passing why the Caen researchers would not have discovered high-temperature superconductivity even if they had made their measurements at low temperatures: samples, like those from Caen, reheated in air at high temperature are less likely to become insulators, which have the necessary strong electron-lattice coupling. The electron localization, essential according to Chakraverty, is significantly weaker, so lowering the temperature would yield no superconducting transition. IBM head-quarters in the United States was unaware of the dramatic results all summer long.

The news of the discovery of high-temperature superconducting oxides would spread according to a scenario that is classic for any sort of innovation. After a long latency period, with a slow growth in the number of consumers in the know, there follows a short period of extremely rapid growth after which essentially "everyone" knows. Finally, saturation is reached; even the scientific hermits and those who re-fuse to open even a single journal find out what all the rest of the world already knows. What is truly surprising in this case is that the impact of the phenomenon would be enormous well outside science itself. Superconductivity would become theater.

CHAPTER 21

Superconductivity as Theater

The lab directors at IBM headquarters in the United States finally learned about the results obtained at their European branch once the article by Bednorz and Müller appeared in print, a good six months after it was submitted. While they did transmit the news to other laboratories, they attached little importance to it; the days of their pioneering superconducting computer project were long gone. The fact that IBM made no official public announcement of such a discovery was proof for many scientists who had heard rumors about the work that the results were probably not credible.[1] Being skeptical to begin with, IBM's silence only fed their cynicism. Their reasoning was plausible, but succeeding events showed only that cynicism replaces neither intelligence nor the exercise of reason.

At the end of 1986 in all of IBM there was but one researcher who had tried to reproduce the Swiss results, a Californian by the name of Richard Greene. He had started his career working on the superconducting mineral $(SN)_x$ and was thus well positioned to evaluate the new compounds. In fact, for several weeks he was unable to produce the Swiss perovskite, and that reinforced the initial skepticism of the Americans. Whether or not he had good results in the autumn of 1986, the fact is that everyone was making fun of the Swiss work until December 4 in Boston, where the Materials Research Society was holding its annual meeting.

Several thousand specialists in the most diverse aspects of materials and their applications were taking part in a program that included, as usual, sessions on superconductivity. Interest in the field had waned, but, to maintain the meeting's reputation for complete coverage and to avoid offending anyone, these sessions were maintained. Only four speakers were scheduled for the afternoon of Thursday, December 4, one of whom was Koichi Kitazawa, an indication of Japan's industrial power. Kitazawa was a researcher at the University of Tokyo, a member of a group that had begun to investigate once more the family of perovskites $BaPb_{(1-x)}Bi_xO_3$, which had not been popular before. As announced, he spent his time talking about known oxides. The reputation of the group was well established in this field, and its work was considered reliable, but Kitazawa had no earth-shaking results to report.

The session was supposed to conclude a little sleepily, in the usual fashion, with a presentation by Ching-Wu Chu, better known to his colleagues by his adopted American name, Paul Chu. He had been a student of Matthias and had a reputation as a spirited researcher, a bit too quick on the draw. In the middle of the 1970s he had accepted the wild suggestion that copper chloride became superconducting at low temperature and high pressure. Upon returning to Russia in 1978 after a stay in Chu's lab, a Russian, Alexander Rusakov, even maintained that copper chloride became superconducting at 200K, a very high temperature.

Like Kitizawa, Chu spoke about superconducting oxides for the prescribed twenty minutes. At the end, as if in passing, he mentioned that he could confirm the findings of a Swiss group in Zurich that had found superconducting oxides at 30K. He had been able to reproduce the compound even if he did not yet know its exact composition. Finally something new to chew on, and a high T_c to boot! The audience suddenly woke up and started asking questions. Someone became quite agitated and, abruptly starting to speak, reached the podium and straightaway took over the transparency projector. It was none other than Kitazawa, suddenly much more voluble than during his entire talk, showing two transparencies that had been meticulously prepared. His group was also working on the Bednorz and Müller oxides. They confirmed the high T_c, they had measured the Meissner effect, and they were just about to pinpoint the structure of the superconducting phase. The participants were convinced that Kitazawa had spoken up only so that Chu would have to share the spotlight. In fact, however, before this session began, Kitazawa had obtained permission to speak again the next day on the same subject, at the opening of the session. At 4 A.M. he called his lab to report the news and to hear about their latest results: the structure of the superconducting phase had been determined. On Friday, as scheduled, he recounted in detail before a particularly attentive audience the work carried out at the University of Tokyo in Shoji Tanaka's laboratory. The superconducting phase had a perovskite structure with the formula $Ba_2La_2CuO_4$.

The Japanese group had not gotten involved in this work by chance. Since 1980 Shoji Tanaka and two colleagues had studied the compound $BaPb_{(1-x)}Bi_xO_3$ with the highest T_c and showed that it had a very low electron density.[2] What's more, a scientist from the Tokyo group, M. Takashige, had arrived in Zurich during the summer of 1986 and had immediately started X-ray diffraction studies to determine the structure of the superconducting phase. In Zurich, as in Tokyo, the Meissner effect in BaLaCuO had indeed been observed, an unambiguous signature of superconductivity. With this double confirmation of the results of Bednorz and Müller, the large American laboratories, until now on the sidelines, got heavily involved in research on the new superconductors.

It was only in January that the fever would hit Europe. For Alex Müller his period of tranquillity, which he knew could not last indefinitely, was over. With the Boston meeting European predominance in this tiny domain of physics also came to a halt. Kitazawa and Chu had just started the race for new superconductors.

The sociology of these events is instructive. While the discovery took place in Europe, a conference in the United States was required before knowledge of the discovery became widespread, a mark of the eminent position that the United States continues to enjoy. In addition to the Swiss, the actors in the drama were, on the one hand, the Japanese and, on the other, Paul Chu's research group, composed of immigrants or sons of immigrants, mostly from mainland China, where Chu had visited in 1979 to create a research group on superconductivity. Chu's group, much less powerful than the large companies, would set a frantic pace for research in the revived field.

Paul Chu was born in 1941 in southwest China. His parents were nationalists, and his family moved to Taiwan in 1949. Chu did his early studies in physics in Taiwan and arrived in the United States to work on his master's degree at Fordham University in New York City. He received his Ph.D. degree in San Diego with Matthias and, after a brief stay at Bell Labs, he became an assistant professor at Cleveland State University and then professor at the University of Houston in 1979.[3]

Chu noticed the article by Bednorz and Müller at the beginning of November 1986 and, with his group, already had many results that he didn't discuss in his talk. He had decided not to follow the Swiss recipe for making the compound but, rather, to mix powders of the oxides and carbonates of the elements required (La_2O_3, CuO, $BaCO_3$). He would then compress them to get discs about a millimeter thick and heat them to the temperatures indicated. The method was completely successful and considerably simplified the preparation of large numbers of samples. Grinding powders is an ancient technique. To compact them and heat them to high temperatures is to carry out processes as old as the fabrication of porcelain, in which China has a very rich tradition. Chu's other trump card was his ability to measure the properties of BaLaCuO under pressure. Few experimenters were able to reach high pressures at low temperatures and measure electrical properties under these conditions; the experimental setup was delicate. Chu maintained a permanent arrangement for carrying out such measurements, because the data are so useful; when a crystal is deformed, its phonon spectrum and electronic properties are changed. The result of his experiments surpassed every expectation: he obtained a critical temperature of 40K.

An unrepentant optimist like Chu could draw two conclusions that are not totally independent. The first was that, since high pressures had such a big effect, simply substituting bigger atoms for certain atoms of the compound should yield the same effect, since the large atoms would have to compress the crystalline structure to make a place for themselves. Raveau had not reasoned any differently when he followed up on the work of the Bordeaux group to make his oxygen detector. In other words, Chu's first experiments led him to think that it was possible to modify the chemical formula to obtain higher T_c superconductors. His other conclusion followed from the first: it was conceivable that there might be a whole family of compounds like $BaPb_{(1-x)}Bi_xO_3$ that were superconducting. This would stand in stark

contrast with organic superconductors, where even today the number of superconducting molecules is tiny compared with the huge number of organic macromolecules. Finally, the results of Chu's high-pressure experiments also suggested that
Bednorz and Müller had not happened upon the family with the highest critical
temperature, if indeed such a family of compounds existed.

Chu put his all at risk right from the start of his experiments. This is not surprising because, according to R. Hazen, a mineralogist who would soon be called to
his rescue, Chu had decided during the summer of 1986 "to abandon superconductors altogether if he couldn't achieve 30K by 1989, the year of his next contract
renewal."[4] If this is true, Chu looks like the Rastignac of physics, a country bump-
kin from Balzac who comes to Paris and quickly strikes it rich. From the end of November, Chu had his gaze fixed on the Nobel Prize and on money, two objectives
that are not at odds with each other but at the same time do not necessarily point in
the same direction. There is an issue here that goes beyond Chu himself. His determination was in part a reflection of the problems faced by American university laboratories to ensure their funding. Chu, at the time a temporary program director
at the National Science Foundation, had a clear impression of how hard it was for a
group to get funded for research in a relatively stagnant area. The leap in the record
for the critical temperature looked like an unexpected opportunity that he had to
seize, a strong second wind for all those who had known the golden age of superconductivity and felt that the important events in physics were not happening in
their laboratories. With all this, it isn't very surprising that the superconductor revolution took quite a different turn with its transplantation to the United States.

After the Boston conference Kitazawa was invited to Bell Labs and to Stanford,
to the laboratory of Ted Geballe, one of the earliest students of Matthias. In the
following week Kitazawa received by fax a draft of an article containing all the results of the Tanaka group, and he soon had copies in circulation all over the United
States. As for Chu, right after Boston he met up with one of his former students,
M. K. Wu of the University of Alabama. Once Wu had been brought up to date on
recent events, the two decided to collaborate. Chu was well aware that, if he wanted
lead a parade, he would have to march with others, since his group was too small;
Wu was just the right collaborator.

Everything was now in place. Within a few weeks the search for a fundamental
explanation of high-temperature superconductivity would be reduced to a competition to synthesize the compound with the highest possible temperature. Success in
the competition would be read as a measure of the relative excellence of the laboratories of the major industrial countries, indeed, of the countries themselves. The
year 1986 was not over before reports appeared simultaneously from the University
of Tokyo and AT&T Bell Labs with the news that the substitution of strontium for
barium in the Bednorz and Müller oxides yielded a critical temperature of 40K.

A third institution also announced precisely the same result: the Institute of
Physics in Beijing, a laboratory not exactly at the center of high technology. In real-

ity, making new compounds required rather modest equipment, well within the possibilities of even poorly equipped laboratories, provided they had at their disposal the right powders and abundant manpower to prepare the samples. For the first time in the history of physics, poor countries could play the game on almost equal terms with the rich. To stay involved in the field even when the cost of the experimental equipment proved too much for them, some of these groups adopted the unusual strategy of volunteering to publish the articles flooding publishers and referees. This was the case, for example, with the National Physics Laboratory of New Delhi. Beyond carrying out research on new materials, it helped spread the innumerable publications in which the authors vaunted their merits.

The initial euphoria soon gave way to the bitterness of a competition in which the only thing that counted was the announcement of a higher critical temperature, together with an approximate chemical formula. On December 27 the *People's Daily* of Beijing announced on page 1 the extraordinary properties of BaLaCuO and even envisaged that superconductivity might exist at 70K. A curious coincidence: Chu's group in Houston had obtained results approaching this. Chu escalated the media activity by convoking a press conference for December 30 in Houston, claiming the world record for superconductivity at 52.5K. On December 31 the *New York Times* published this result in conjunction with previously unknown results from Bell Labs, 36.5K for SrLaCuO. The breakthrough by Bednorz and Müller was hardly mentioned, no doubt because, even though they were from IBM, they had the poor taste to work outside the United States.

In fact, the Bell Labs group had just submitted an article to *Physical Review Letters*, two weeks after Chu. The two articles were, nevertheless, published in the same issue, dated January 27, 1987.[5] (*Physical Review Letters* had acted similarly with two articles on flux quantization in 1961.) Each paper included a bare minimum of theoretical discussion to show that the authors knew what they were talking about. Here also lay the roots of a debate that would gradually become more heated. Chu and his collaborators leaned in favor of a superconductivity at the interfaces between the grains that made up the ceramic. R. J. Cava and his Bell Labs colleagues, however, after examining its properties in a magnetic field, were more conservative. The usual interaction between the electrons via the phonons was sufficient, they felt, to explain the properties of these new compounds. Apparently benefiting from a rereading of the proofs of his article, Chu added the following: "Note added, 6 January 1987: Detailed examination of the results on the sample in Ref. 2 indicated that the sample exhibited a critical temperature of around 70K and that the sharp resistance drop occurred at around 60K, although the zero resistance state was not reached. We, in collaboration with M. K. Wu at the University of Alabama at Huntsville, also found that the replacement of Ba by Sr produces a critical temperature of around 42K at ambient pressure."

The "Ref. 2" Chu was referring to also mentioned that the sample that had yielded a critical temperature higher than 56K had been destroyed and couldn't be reproduced. This sample had been produced by Wu's Alabama laboratory, which thus was clearly a real partner in the group and not just "an extension of the Houston laboratory."[6] Afterward, the members of the two groups each had their own version of what happened at this time. According to Hazen, who faithfully relates the events as they were seen from Houston, Chu had a long meeting with Wu and one of his students, Jim Ashburn, in the beginning of January. They decided to see what would happen if the atoms of lanthanum were replaced by neighboring rare earth atoms: ytterbium (Yb, with atomic number 70), lutetium (Lu, atomic number 71), and a third, located in the third column of Mendeleyev's chart just above lanthanum, yttrium (Y, atomic number 39). Yttrium is much lighter but slightly smaller than lanthanum. All this makes sense. Because the replacement of barium by strontium had already been done, Chu's strategy was to try another trick, replacing lanthanum by something similar but smaller, yttrium. For all those in the race there was no question of touching the copper, since its presence, for one thing, and its atomic environment, for another, seemed indeed to be key to the new superconductors.

Between the proposal of a project and its realization, however, there is always some distance. Chu had delegated the unpleasant task of grinding and mixing the powders to two undergraduate students. Apparently, they didn't do very well, and they never managed to produce a superconducting sample, even starting from the elements of the compound discovered by Bednorz and Müller.[7] During this time the Alabama people continued to vary the composition of the SrLaCuO compounds because they had no yttrium at their disposal, and it had been decided that Chu would order it. Nevertheless, at the suggestion of his university, Chu filed a generic patent application for all the superconducting compounds that he could imagine, including those he had never fabricated. For some, such as yttrium, he had no idea how they would behave. Wu was apprised of the existence of this patent application. Chu's general objective was not understanding but inventing new materials, and he wanted his inventions to be covered by a patent. Others, no more interested in understanding than Chu was, might well have been spying.

The *People's Daily* returned to the subject on January 17 to say that superconductivity at 70K was unstable but that a LuBaSrCu compound was stable at 40K. Given this report that could have been written in Houston, Chu became paranoid about espionage not only by Beijing but also by his American colleagues. Apart from several coincidences that might be considered suggestive but certainly not proof, however, it had to be demonstrated that spies were or were not observing his work.[8] In a milieu in which communication takes place more and more via noninstitutional channels, it is excessive to bring up espionage right away. The spy is often a colleague who is off on the same trail as you are. The intellectual breakthrough was the achievement of Bednorz and Müller; to go beyond that was more a question of seizing the opportunity rather than having exceptional prowess. The race was bitter, because those who threw themselves into it were first of all convinced that they were approaching critical temperatures equivalent to or even higher than those of liquid nitrogen at atmospheric pressure (77K; liquid nitrogen is 3–4 times less costly than liquid helium in Europe). This explains why, in most centers where researchers got started on this track during the first part of 1987, scientists who believed they were onto something that no one else knew were often paranoid about it, even if it wasn't a revolutionary compound. Such an experimenter would suddenly adopt the habit of locking his desk before he stepped a few meters away from it, or he might abruptly treat all his colleagues with royal scorn. The laboratory might suddenly divide into two camps; it was a miniature cold war. In France this infantile "everyone for himself" attitude and the accompanying tendency to keep information secret delayed meetings about coordinating efforts on the new superconductors for several months.

Starting from their results on compounds with strontium, as well as more recent ones with calcium that produced superconductors with lower critical temperatures, Wu, Ashburn, and another thesis student, Chuan-Jue Torng, tried a less empirical approach to finding the optimal substitution. Without changing the structure in the neighborhood of the copper, this would yield the highest critical temperature. On January 17 Ashburn tried to correlate all their results with the radii of the atoms involved; he reached the conclusion that he had to start from the composition $Y_{1.2}Ba_{0.8}CuO_4$. Wu had no yttrium, but happily his colleague at the Huntsville space center provided some for him. On January 28 the powders were put in the oven for the night.

On January 29 around 5 P.M. the Huntsville group started measuring the resistance as a function of temperature.[9] At 93K it dropped precipitously: the compound became superconducting well above the temperature of liquid nitrogen. For years discussion about applications of superconductivity had been tempered by the remark that life would be so much simpler if we had superconductors above liquid hydrogen temperatures. Here was one well above the temperature of even liquid nitrogen, the most common of the liquefied gases, and chemically inert as well. Wu, who is much less an extrovert than Chu, recounts how his hands trembled for a half-hour.[10] It was only later that he told Chu, who didn't even know that his partners had already started working on yttrium compounds: "We hit the jackpot!"[11]

The next morning Wu took a plane to Houston carrying his sample, which he hoped would not lose its superconducting properties during the night, as so often happened with this kind of material. In Houston, on January 30, a measurement of the resistivity of the sample of YBaCuO reproduced the results of the evening before; it was indeed superconducting at 93K and thus stable. Chu had just won the race with two samples made in Alabama — one with strontium and the other with yttrium — more than 40K higher than the old record.

The composition of the miracle sample was one of those mentioned in the catalog that Chu had included in his patent application, but this was only the first step for someone wanting to bank the profits. He would also have to identify what was superconducting in the dense mixture of little green grains, translucent like a wine bottle, and little black grains that were perfectly opaque. Wu and Chu each had their own ideas about why they should try yttrium, but a fundamental premise for both was that the structure of the region surrounding the copper should stay the same as it was in the Swiss compound. The new compound obviously violated this premise: the two kinds of grains meant there were two phases with different structures, and the original structure was not conserved.

Chu proposed to a group of crystallographers housed in his building that they investigate the structure of the new compound with X rays; Ken Forster, a young thesis student from this group, was put in charge. Since the samples were polyphasic, nothing precise could be determined from the X-ray absorption spectra unless Chu furnished the chemical composition of the original sample. Chu refused, and Forster's advisor ordered him to stop working for Chu. The paper describing the electrical resistivity measurements (and the susceptibility measurements that confirmed the Meissner effect) thus had to be submitted without any structural information. The fear of being deprived of the rewards of his discovery pushed Chu to submit two articles immediately to *Physical Review Letters*. The one described the high-pressure measurements that provided the justification for the idea of using yttrium. The other presented the results obtained with the Alabama sample. Everyone who had participated in the race was an author. Chu, the leader of the enterprise, was last on the list, in contrast with the previous article on BaLaCuO, for which he was first. This time, however, the first reference was to his patent application, for which he was the only author.

What followed is even more astonishing: Chu insisted that the (second) article be accepted without being submitted to referees. The editor of *Physical Review Letters*, Myron Strongin, refused. Chu then proposed that the article be refereed with the chemical formula masked, and again Strongin refused. Finally, Chu submitted his article with a formula included; in return, Strongin and he agreed on two referees in whom he had complete confidence. Strongin was bending the rules; the names of the referees are normally kept strictly confidential (which doesn't mean that authors don't sometimes make a pretty good guess). The articles were received on February 6, accepted on February 10, and sent to the printers. Preprints, copies of the paper that are often distributed before publication to attract the attention of interested colleagues, were not sent out; even the coauthors did not receive a copy.

Rumors about a new record temperature nevertheless spread around — there were leaks. Curiously enough, the rumors mentioned ytterbium (Yb) and not yttrium (Y) as one of the elements of the new compound. This switch wasn't as bizarre as it sounds: all copies submitted to the journal contained this error more than twenty times, each time the Y was supposed to appear. There was also another "error": the 4 in the chemical formula $Y_{1.2}Ba_{0.8}CuO_4$ was replaced by a 1. It is difficult to believe that these errors weren't deliberate. When researchers who got wind of these rumors telephoned Chu to find out if this story about an ytterbium compound was correct, he didn't deny it. Whatever one might think about those who might exploit this information, Chu's refusal to deny the rumor amounts to misleading his colleagues. In the patent battle Chu had made several major advances; this was war, after all. It is hard, however, to reconcile his behavior with the respect that the scientific community still attaches to the Nobel Prize. The irony of the story is that only six weeks later several large laboratories showed that BaYbCuO compounds were themselves superconductors at temperatures above 90K.

Robert Hazen has identified the origin of the leak. It was a collaborator at Physical Review, a physicist at Brookhaven, who during a meal with one of his students recounted Chu's demands as a good joke. The girlfriend of the student worked at a local newspaper on Long Island, where the laboratory, an old military base from World War II, is located. She was writing an article on *Physical Review Letters*; what she learned from her friend that evening made her piece a good deal more gripping. Hazen exonerates Chu with the excuse that his behavior was necessary to protect the fruits of his labor, a claim that to us seems to cheapen the traditional value attached to the quest for truth. Chu corrected his error only at the very last moment, on February 18, in a telephone call to Strongin in which he provided the correct formula for the compound. The referees were informed, the proofs were corrected, and the article appeared without error. Two days earlier Chu had held a press conference after the annual meeting of the board of the university; he announced the discovery of a superconductor at 93K without revealing the formula before his article appeared in print.

During this entire period Chu's group, with little competence in crystallography, had not succeeded in deciphering the structure of the compound. Through a Chinese-American colleague Chu made contact with Hazen, a geologist from the Carnegie Institution. Hazen was a total outsider to the world of superconductivity, yet he was knowledgeable about perovskites from his geological activities. The crystals of the two phases, the green and the black, were both typically of order one micron, a millionth of a meter, in size, at the lower limit of reliable operation of his apparatus. All the crystallographers had run up against the same difficulties. In addition, Hazen had the misfortune of starting to work on the microcrystals of the green phase, which would not turn out to be superconducting. Only later did he work on the black phase, but, in the meantime, others beat him to it.

Bellcore, a company formed in the breakup of AT&T that later became Lucent Technologies, and other large laboratories confirmed the results of the Wu-Chu collaboration quite early on, but, by a stroke of good luck for Wu and Chu, only after their paper appeared. A French chemist employed by Bellcore, Jean-Marie Tarascon, had prepared samples with yttrium on January 3, but no measurements were carried out right away. When his laboratory heard the rumors about yttrium — around the time that the paper was published — it was only a question of hours.

Excitement took hold of the world of condensed-matter physics, including Europe; Bednorz and Müller had nothing to do with it. An enormous number of papers flowed into *Physical Review Letters*. Each author touted his merits, as if a club with limited membership were just getting started. It was only too evident that in such a climate of hypercompetition what was important was the announcement of a new result; evidence and backup experiments were considered unnecessary. Everyone knew that the results obtained would be out of date a few weeks later. By the beginning of March the major American weeklies had all covered the story in detail, and *Time* magazine made the new superconductors its cover story. As Hazen recognized clearly one year later: "We had found a structure model that explained our data reasonably well, so we were right in that sense. But with better crystals, or more time, we'd have obtained better data and a better model. And no matter how well we did, someone else was going to do it better next month, and then better next year, and so on and so forth. So, what was right?" [12]

The liberal and humanistic scientific tradition, if ever it existed except in the romantic illusions of young scientists, was ridiculed. The scientists were no longer wise men but, rather, merchants in the temple. Surely, the circumstances were exceptional, but is it not in such exceptional moments that the true nature of a social phenomenon is revealed? It is impossible to see in this singular situation anything at all related to the quest for fundamental understanding; the breakdown of the liberal model of scientific activity was manifest. But the worst was yet to come, soon, at the 1987 spring meeting of the American Physical Society at the Hilton hotel in New York City.

These conferences generally include plenary sessions in areas in which recent progress has been particularly significant, specialized colloquia, and a number of parallel sessions with short contributed talks that are mostly of routine interest. Wide corridors are particularly important for renewing contacts, for gossip, and for making one's virtues known to potential employers. The latter was of special interest at the end of the 1980s, since the U.S. economy no longer was the undisputed world leader. The organizing committee had planned, of course, to devote a session to the new superconductors; Alex Müller was the opening speaker, followed by those who had successfully followed the paths he had cleared, Tanaka and Chu. The situation with the contributed papers, however, was more complicated. Each large laboratory wanted to have its say, more to affirm that it was part of the game than to contribute to a real scientific debate. Twenty-five groups asked to be heard, a sign that pointed toward a marathon session not at all in the usual relaxed style. Press releases carried more weight than the scientific presentations. The usual participants in such meetings were reinforced by journalists, television teams, and a fair minority of researchers from abroad who had come to make up their minds about whether there was any truth at the bottom of all the most fantastic rumors that had been spreading.

The meeting room, modestly named the "Rendez-vous Trianon," was overwhelmed from the moment the doors opened; a huge crowd had been waiting over an hour, and there were only twelve hundred seats. In fact, this was not the Trianon but, rather, the Bastille, one July 14 evening. In the pushing and shoving outside the room, several people were hurt. Some raced up the down escalator to avoid being crushed by the crowd on the lower level; opening supplementary rooms with TV monitors provided little relief. The atmosphere was electric, to such a point that the chair, at the insistence of the fire chief, threatened to interrupt the session if the emergency exits and corridors were not cleared. Finally, the meeting started at half past seven.

Müller surprised his audience at the end of his opening talk by showing one last transparency illustrating the structure of Chu's black phase, which had been determined by a group from IBM. (At this point Hazen had the right result, but there were three other groups that had also reached almost the same conclusions about the structure.) Then came Tanaka, Chu, Zhong-Xian Zhao from Beijing, and Bertram Batlogg from Bell, an old hand at superconductivity. Picking up the tradition of Dewar's public demonstrations of liquid helium, he got right to the show; he was the first to demonstrate new superconductors, a doughnut-shaped winding of 2.5 cm as well as a flexible cable. The two objects were useless except to suggest that applications were right at hand and to underline the differences between the capabilities of the university groups and Bell Labs, something no one doubted, even though Bell was not as strong as it once had been. Interrupted from time to time by breaks, the session lasted into the middle of the night with little weakening of the spectators' enthusiasm. No one who participated has forgotten it. The spirit was so unique that one physicist from Bell talked about the "Woodstock of physics," alluding to the famous music festival of the 1960s. That expression was picked up the next morning by a journalist from the *New York Times*. Because most everyone knows at least roughly what Woodstock was about, and not at all about superconductivity, this unexpected image, the Woodstock of physics, counted enormously in the success of the new superconductors in the media. It was repeated by all the popular dailies, and it continued to please the physicists as well.

Expressions have nostalgic weight. Woodstock was a happening that no one could have ignored, whether or not they liked popular music. For the physicists it meant renewed contact with society, a new coat of paint for their image. Woodstock carries a convivial image in contrast to all the applications of physics that are perceived as menacing, the triumph of machines, big science, arms of massive destruction, and the nuclear industry. Amil Nidura, in *Physics Today*, was one of the rare commentators to question the significance of this meeting of the American Physical Society.

The new discoveries resonated enormously with the public, or, more precisely, with the enlightened middle class in the industrialized countries. The absence of major international conflicts polarizing the attention of the media was no doubt largely responsible — it is likely that several years later the resonance would not have been nearly so strong. The transformations that the Europeans adjusted to in the succeeding years, such as the fall of the Berlin Wall, were much more important for this generation than the "revolution" in superconductors.

So many physicists who had never even gingerly tested the waters of superconductivity now jumped head first into the maelstrom. Everyone marvels at how the

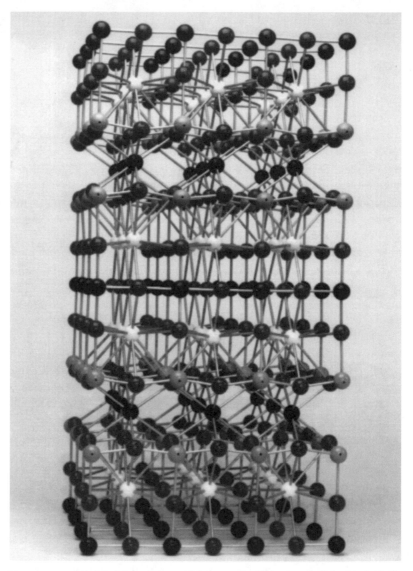

Perovskite structure. Model of the structure of the perovskite YBa$_2$Cu$_3$O$_7$ (YBACUO) deficient in oxygen. Gray: oxygen; dark gray: copper; white: barium; black: yttrium.

Japanese turned ceramics kilns usually reserved for the clay chefs-d'oeuvre of children into ovens for cooking up samples of YBACUO. Others, Nobels in superconductivity among them, found it suddenly imperative to immerse themselves in a field that they had long disdained. It's hard to avoid viewing such intellectual ferment, with so many physicists abruptly changing course, as a sign not so much of flexibility and renewed vigor as of lack of interest in the myriad microspecialities that nourished daily life in condensed-matter physics. Even if you had scarcely any interest whatever in taking part in this superficial race, you were asked to join one or another experiment or collaboration. Those who had not bothered for years to say one word about superconductivity, leaving pots full of problems that students knew nothing about, suddenly found it urgent to go back and teach just what had been taught twenty years previously.

Many of the properties of the new superconductors, however, involved problems that had not yet been solved. They were anisotropic, their coherence length was short, less than about twenty angstroms or several interatomic distances, exactly like certain layered compounds from the 1960s that were no longer fashionable. The contribution of the electrons to the specific heat was very large, like heavy-fermion superconductors that no one understood clearly. But these old subjects, some of which were perfectly well understood, became the last word in scientific inventiveness, even when their applications to the new superconductors were rigorously the same as for the old.

It wasn't just fashion. This time, in addition to the customary lemming instinct, justification would come again from the miraculous applications. Once more we would be able to look forward to a free lunch. We would transport energy without losses, as if it didn't take any energy to liquefy a gas. We'd have a miracle in electronics, as if all the computer manufacturers, IBM included, were not watching their profits decline as the demand for large systems collapsed. In fact, the government did open the money bags a bit to fund the new superconductors but only for a short time, to the detriment of other scientific activities, and — one time does not a habit make — even to the detriment of big scientific equipment. The American high-energy physicists, who were having great trouble getting funding for their future biggest accelerator in the world, tried to save their project by explaining that the new superconductors would reduce the operating costs of the 84 km long machine. There were few discordant voices. One of the rare ones was heard at a July 1987 European Economic Community scientific meeting in Genoa with the goal of coordinating efforts in superconductivity. This was a representative of the Siemens company who explained that, even under the most optimistic performance scenario at liquid nitrogen temperatures, superconducting cable had no obvious economic advantages:

> In our estimation of costs and losses, we have considered cables with alternating currents at 345 and 380 kilovolts carrying 5000 megawatts of power. Previous studies have demonstrated that this is the threshold at which superconducting cables become advantageous. The most important savings come from the refrigeration system; the cryogenics envelope might yield a gain of 33%. Substantial savings is expected in the cost of the refrigerator, which will go from the present 9.1% of the installation price to 3.6%; what's more, no helium will be needed. In sum, the retail price of installing a cable at 77K is 75% of that for a cable at 4K. In addition, a remarkable reduction in losses has been estimated, from 130 kW/km to 40 kW/km. It remains to be shown in the future,

however, whether these advantages are sufficient for a 77K cable to be competitive with the most highly developed conventional cables.[13]

The mechanical properties of these new compounds were just the opposite of what would be desirable for cables. A considerable scientific and technical development effort certainly seemed necessary, and it had yet to be demonstrated that the benefits would be worth the costs. Nevertheless, Siemens got heavily involved in developmental research on the new superconductors largely subsidized by public funds from Germany.

Understanding the mechanism by which these compounds became superconducting at high temperature was and still is a real scientific problem. Structural studies ultimately yielded a clear image of the superconducting material. The final compound obtained varied significantly with the different recipes for cooking the initial ingredients. Differences in oxygen content could strongly influence the properties of the material and even make superconductivity disappear. Everyone quickly agreed that the superconductivity was produced in CuO_2 layers — again, as in the 1970s, it was thus a problem of anisotropic superconductivity. A number of physicists, among them some who had been active in the 1970s, didn't bother to look back over the work that had already been done then and started anew. They carried out all the old experiments on the new samples; the results, which didn't depend on the critical temperature, were completely predictable. Many properties of the new materials were studied in this same fashion; obstacles were found at every step.

The journals were crushed under the weight of papers waiting to be published; the classic journals could no longer keep up with them, even when they added considerably to the number of pages per volume. Because subscription prices had been set at the beginning of the year, the journals were losing money, aggravating their financial situation, which had never been robust. New journals were created with much higher subscription rates and a guarantee of publication within a reasonable time. In fact, however, much of the information traveled by telephone or computer without review. Only the reputation of the authors provided any indication of the likely reliability of the result.

Real scientific communication was becoming improbable in the face of the thousands of publications in this small field of physics, which had reached the boiling point. What mattered during the years 1987 to 1990 was a priori opinion. The results of the quest for materials produced, not for the first time in this field, a great number of bizarre announcements. But alongside these reports, not all of which were in good faith, the number of compounds that are superconducting above liquid nitrogen temperatures grew, as did the record temperature, which was established around 138K in mid-1993. The principal milestones in this increase in temperature are presented in table 21.1. Things became more complicated, however, with the discovery of a superconductor without copper, with a T_c of the same order of magnitude.

TABLE 21.1

Highest Critical Temperatures and Their Corresponding Compounds, 1986–1993

YEAR	COMPOUND	T_c
1986	$La_{1.8}Sr_{0.2}CuO_4$	38K
1987	$YBa_2Cu_3O_{7-\varepsilon}$	93K
1988	$(Bi/Pb)_2Sr_2Ca_2Cu_3O_{10+\varepsilon}$	110K
1991	$Tl_{1.6}Ba_2Ca_2Cu_3O_{10-\varepsilon}$	128K
1993	$HgBa_2Ca_2Cu_3O_{8+\varepsilon}$	138K

Note: The quantity ε is a very small fraction of unity.

So many new superconductors appeared that the question of a possible new type of coupling between the electrons arose, even though the usual kinds of electron pairs were responsible for the phenomenon. This might seem strange to nonphysicists, but the surprise of suddenly finding, just like that, a rich class of new high-temperature superconductors gave some theorists pause. All their notions and explanations had been based on superconductivity at low temperature, where it was well understood. Maybe they had missed something important that simple, overly rough approximations had neglected, something that would become more important at high temperatures. And so they asked whether the physics of solids, which works so well at low temperatures, was still valid at higher temperatures. Physics is not mathematics, in which objects are constructed as elements of a theory. Discoveries in physics of relevant phenomena and the most pertinent materials do not necessarily occur in a logical order. Physics lives always with a suspicion that the general synthesis that it proposes at a given moment may be less general than the common opinion of physicists admits.

The year 1987 finished with the awarding of the Nobel Prize. Given such an explosion of results, it was almost obvious that the field would be superconductivity, but who would get it? The response of the committee was unambiguous: Bednorz and Müller were the only recipients.

After a year so rich in unusual events, the search for superconductors at high temperatures, this new kind of war against the cold, could only take on a less chaotic turn. As could have been predicted, the applications of superconductivity have not changed much since then. A limited number of electronic components have been developed. Generally, however, applications of superconductivity today are neither more nor less successful than they were previously, in spite of the leap of 100K in critical temperatures. The race for new materials slowed down at the beginning of the 1990s. In spite of everything, the search for new superconductors still has an ambitious goal, one that would have seemed completely unrealistic in 1986 but which now no longer seems completely improbable. Even the specialists, in this field always more conservative than the futurists, who generally have

The Nobel Prizewinners. J. Georg Bednorz (*left*) and K. Alex Müller congratulate each other just after the announcement of their Nobel Prizes.

confidence in progress, accept this. The goal is finding room-temperature super-conductors.

In the long term we can note only that this cold war we speak of, which was so frequently tied up in the worst dramas of the century, is limited today to a competition for prestige and technical dominance. These two objectives have always been the raisons d'être of science policies and many physicists. We can also note, however, that what was yesterday the activity of small groups in the context of economic expansion is today real life for thousands of specialists battling for their professional survival.

In the future no more than in the past, obtaining interesting materials has little chance of occurring as the result of a prediction. What's at stake for theory, in fact, goes well beyond room-temperature superconductivity. If the old theory remains valid and the new superconductors are interpretable more or less in terms of this theory, there will be no possibility for a break in the paradigm. This is presently the point of view of most physicists working in the field. In this case it must be recognized that the sole motivation for supporting the multitude of researchers is pure empiricism, since it is arguments based on chemical structure that are the tools of progress. If this is indeed the case, what is called, sometimes abusively, the physics of materials would no longer be a living branch of physics, one that aims to deepen its fundamental concepts and tries to propose a philosophy of nature. Even the concept of the physics of materials would then become an oxymoron: materials would be invented based on physics ideas that are unchanged. This would be an art of materials in the same sense as an art of engineering. It would be a purely technical development, but this wouldn't mean that from time to time a physicist would not come by to find out what's happening. In fact, this evolution is already under way.

Nevertheless, in the craze that has thrown so many physicists, humble and illustrious, into a race for new superconductors, we can discern some signs of hope: the hope that the "revolution" of superconductors would be a revolution of epistemology; the hope that this cold war that is the search for new superconductors would carry in it the germ of an epistemological crisis from which would issue a new synthesis of our knowledge, of a reorganization of our understanding of a solid. It's an extremely optimistic vision. If such an enterprise was crowned with success, not only would condensed-matter physics draw from it a new legitimacy, but it would also take over an intellectual field that now belongs to the chemists. More modestly, it is not impossible that the "theoretical" understanding of these materials would become a purely technical activity based on well-established methods, something like the calculation of band structure in semiconductors or molecular orbitals in chemistry.

CHAPTER 22

Almost Twenty Years Later

Readers of Alexandre Dumas will remember that *The Three Musketeers* was so successful that he acceded to popular demand and provided a sequel, *Twenty Years Later*, with the same characters. Despite the weight of years, in it the musketeers had succeeded in society and were just as flamboyant as ever. The story is not the same for the superconductors that some still call new, as in fact they approach adulthood. Their family turns out to be a much larger and more diversified bunch than seemed possible just after their discovery. Nevertheless, the child prodigy that would be superconducting at room temperature is not among them, and the promised applications may arrive only with the wisdom of age.

Understanding high-temperature superconductivity remains a problem. The original motivation for synthesizing the dozens of superconducting oxides that appeared within a few years was to explore in detail the ones that contained copper, since it was recognized early on that interactions in the CuO planes were a crucial determinant of high T_c. The succeeding development of ceramics that contain no copper but are still anisotropic has generalized the problem. Like everyone in this field, we have little schemes in our head that, by their power of suggestion or the elegance of the representation that underlies them, at least temporarily become our preferred "explanation" of high-temperature superconductivity. The elegance of these ideas, however real, does not mean that they are the key to solving the problem. The history of traditional superconductivity clearly demonstrates that, if the theory of Bardeen, Cooper, and Schrieffer had been formulated elegantly, many fewer theoretical papers would have been published afterward. Thus, it seems reasonable to pay attention to the words of an expert whom we have met before, Philip Anderson: "The standard preamble for all kinds of papers on theory (or experiment, for that matter) in the field of high T_c usually contains the phrase 'since there is no consensus on the cause of high T_c superconductivity' or words to that effect, and often proceeds to justify thereby yet another implausible conjecture as to some aspect of the phenomenon."[1]

We have to take this judgment seriously. It comes from a Nobel Prize winner who recognizes that for five years he himself had defended a theory that he claimed was a priori beyond criticism because of its generality but which "turns out to be one of those which must be consigned to the dustbin."

This said, we must not confuse cynicism with intelligence about this subject. It would be cynical to remark that there have been roughly ten thousand publications on the theory of high T_c superconductors since 1987 (and several hundred thousand experimental publications), and surely few specialists have read even 1 percent of them. Dozens of explanations for the interaction responsible for superconductivity in the cuprates have been proposed, based on the results from all kinds of experiments. It is wise, however, to recognize that beyond these figures a certain number of characteristics of these materials are now well established:

- Electron pairs are still important, but they are different from traditional Cooper pairs. Cuprate pairs have both a short coherence length (several nanometers) and a symmetry reflecting the strongly anisotropic crystalline structure, whereas Cooper pairs are spherically symmetric (and thus independent of crystal structure) and have a coherence length of roughly twenty nanometers.

- The superconducting properties depend sensitively on the quantity of oxygen in the compound relative to the exact chemical formula. One speaks of under- and over-doping in oxygen just as one speaks of under- and over-doping in impurities in semiconductors. Yet the similarity in vocabulary masks fundamental differences in orders of magnitude. In semiconductors the doping is typically one part in a million, whereas in high T_c superconductors it can be as high as several percent.[2]

- The electrical properties of oxide superconductors are very different from those of standard superconductors because the energies of interaction between vortices can be much smaller than thermal energies. Near the critical temperature the regular network of vortices becomes more like a pile of intertwined spaghetti.

Several highly technical characteristics could be added to this list, but they would not invalidate Anderson's disillusioned observation.

While it is possible to speak of a renaissance of interest in superconducting electronics, efforts to find applications of high T_c superconductors have generally achieved meager results. Probably the biggest disappointment is the failure to obtain stable, long—say several hundred meters long—wires with millimeter diameters that can be wound to make a coil. The consequence is that no (high T_c) magnets, transformers, turbines, or motors with much power have seen the light of day. The longest high T_c wire is about a dozen meters long, and the price per meter is exorbitant.

The picture is somewhat different in the domain of low-current applications (i.e., superconducting electronics), which benefit from the progress in instrumentation for microelectronic circuits. "Foundries" have been established that make superconducting circuits to order, just like semiconductor microprocessors, which have been available this way for years. Josephson junctions based on oxide superconductors have been on the market for more than ten years, even though the success rate in fabrication is currently of the order of 10 percent. SQUIDS are also available, for the portable magnetometers useful in geophysics, for example. Yet their sensitivity is an order of magnitude less than their classical counterparts that function with liquid helium. A number of logical circuits have been tried, but, when economic realities are taken into consideration, the only seemingly viable avenue for such applications has been superconducting electronic filters for mobile telephones. These filters have been installed in relay stations for the U.S. mobile phone network, in which they make it possible to increase greatly the number of communication channels in a given frequency band. Ironically, however, because the filters operate without problems for a long time, the market for replacement filters is unlikely to be very large. Nevertheless, a number of groups are continuing to explore the development of various superconducting devices, especially those that permit high-speed communications at 100 Gbit/s. It would be ideal if such devices could be connected directly to optical fibers, but "so far there are no satisfactory solutions."[3]

Federal governments have played an important role in the management of the science and technology related to possible applications of high T_c superconductivity — the end of the cold war did not mean the end of governmental intervention. In the Netherlands (as we have seen with Kamerlingh Onnes) and in other western European countries such as France, England, and Germany, the management of science by the state ("science policy") has been traditional for at least two centuries. Even in the United States, despite all its rhetoric about economic liberalism, the federal government's role in the financing of physics has remained strong.[4] And a new champion of governmental intervention has appeared in the East: Japan. In spite of its economic problems in recent years, Japan has been maintaining a long-term program in superconductivity research with impressive results. The U.S. National Science Foundation and the Defense Advanced Research Projects Agency sent a mission there in the late 1990s.[5] The introduction to the chapter on SQUIDS in the mission report speaks for itself: "In the 1980s, Japan had essentially no superconducting quantum interference device (SQUID) technology. From 1990 to 1996, the Ministry of Trade and Industry (MITI) funded the Superconducting Sensor Laboratory (SSL) to focus on this technology, and by the time of the [panel's] visit to Japan, prototype or product SQUID systems were being made by Daikin, Seiko, Shimadzu, Sumitomo and Yokogawa."[6]

One conclusion is the following: "Overall, although SQUID technology at the component sensor level is more advanced in the United States than in Japan, activity to develop new systems and applications, either LTS (low temperature superconductors) or HTS (high temperature superconductors), appears limited at present [in the United States]. Consequently, if large SQUID markets develop, Japan (and also Europe) are [sic] likely to be strong competitors. This is in contrast to a decade ago, when the United States had a dominant position."[7] Indeed, it is relevant to remark that the history of small U.S. companies involved in superconductivity is the history of small groups of scientists spending much of their time creating start-ups that quickly failed.[8]

Another level of intervention occurs in the management structure of large scientific institutions, which are now trying to adapt the methods of personnel and program management used in industry and supermarkets to the management of scientists and scientific research. Nevertheless, with so many people involved, it is inevitable in spite of these efforts that some of our colleagues who are hardly anarchists but do march to their own drummer will explore some overgrown path outside the managers' purview and develop our knowledge well beyond the planned objectives. Life, even scientific life, is always a little more complicated than one can imagine. Since the French edition of this book appeared, this has been the case for two unexpected materials.

Fullerenes, little atomic edifices similar in form to the domes of Buckminster Fuller, were discovered in ordinary soot. They consist of carbon alone (their chemical formula is C_{60}); the marvelous arrangement of their atoms in crystals was a total surprise. Fullerenes are spontaneously produced when combustion is incomplete — in candle flames, for example, or campfires. The crystals are small and escaped observation until they were isolated in a laboratory; it was years later before anyone noticed that they were superconductors. Their critical temperature can rise to 20K if they are doped with potassium and to more than 30K when doped with other alkaline metals such as rubidium and cesium, but they remain powders, and this limits their possible applications.

Experiments carried out over the last several years at Bell Laboratories suggested that much higher transition temperatures could be reached by increasing the dimensions of the fullerene crystal through doping, but these results could not be duplicated elsewhere, and the original papers have been retracted. The history of superconductivity since the 1940s reveals many instances of poorly interpreted data or unnoticed experimental errors that have led, for example, to unfounded claims of a high T_c for a new compound. The recent publications on fullerenes, however, are different: they appear to be deliberate attempts on the part of one of the authors to falsify or fabricate data.[9] A number of other publications by the same author with various collaborators have also been retracted; each of them, if valid, would have been "a major breakthrough in condensed-matter physics and solid-state devices." A blue-ribbon committee appointed by Lucent Technologies, the managers of Bell Labs, found "compelling evidence of scientific misconduct" by this author in sixteen of the twenty-four cases it examined; the coauthors were "completely cleared of scientific misconduct."[10]

This strange and somehow fascinating story raises a number of questions about the motivation of a scientist deliberately trying to deceive his peers but, more important, about accepted behavior in modern physics. Why were the referees for the famous journals so uncritical? What are the responsibilities of coauthors? Why did the main author's supervisor not look carefully at the data, which often did not exist? The committee noted that this author "demonstrated an ability to write coherent, stimulating papers at a remarkable rate, an average of one paper every 8 days during 2001."[11] Should not such productivity, particularly for an experimentalist, appear extremely suspicious to mature scientists?

Not so long ago, Alan Sokal, a well-regarded theoretical physicist, became famous in a larger community by submitting a paper that sounded scientific but was in fact meaningless to a "postmodern" journal of literary and sociological criticism. The event gave birth to what the media dubbed the "science wars." Now we have an example of physicists and prestigious physics journals falling victim to what can be considered a similar hoax. Many scientists who were more than happy to gloat about the Sokal affair are now embarrassed and silent, as if it were only a family matter.

The story of superconductivity in fullerenes thus provides more plot twists than a spy novel, and a happy ending is not in sight. A new superconductor born in Japan in 2001 looks more practical: MgB_2, magnesium diboride. The announcement of a new superconductor these days is almost as routine as the announcement of a new automobile. This time, however, the experts were awed by the almost biblical simplicity of the chemical formula and the banality of its elementary components—magnesium and boron are very common. The critical temperature of MgB_2 is 38K, well above that for the first of the superconducting oxides that earned the Nobel Prize for Bednorz and Müller. For the ever-narrowing circle of initiates, the real question was not why it is a superconductor. It was, rather, how did Bernd Matthias miss this wonder[12]—Matthias, the iconoclast whom we have credited with an exhaustive exploration of the entire periodic table including ternary compounds. Had he found it, the history of superconductivity would have been transformed.

Why is this new compound so interesting? Its critical temperature does not reach the 77K of liquid nitrogen; it is higher than the temperature of liquid hydrogen, but simple reasons of safety preclude the use of hydrogen as a coolant. Thus, it must rely on liquid helium, just like niobium and other alloys that have been around for forty years. Nevertheless, MgB_2 is preferred for three important reasons. First, it is easy to deal with, since the material is almost perfectly isotropic. After only a few months of effort, one can draw a wire several tens of meters long that is easy to wind because it is coated with iron. In contrast, after fifteen years, making ten meters of YBaCuO still requires a silver matrix. The second factor is its remarkably low price, much lower, it seems, than the price of ordinary copper wire. Finally, the "pulse tube" technique, which eliminates all moving parts at low temperatures in helium liquefiers, makes them as easy to use as home refrigerators.[13] Working in the temperature range of tens of kelvins is thus much less costly than it used to be because the refrigerating power of such devices improves with increasing temperature. Although MgB_2 is not a very high temperature superconductor, it nevertheless seems to open up for the first time the realistic possibility of high-current applications. Estimates that include the cost of refrigeration suggest that a MgB_2 transformer working at 25K will cost about 30 percent less than would a transformer with copper wiring. The situation would then be very different from the forecast of G. Bogner, the Siemens economist we quoted earlier at the dawn of high T_c, who said that, even if the new superconductors worked as well as niobium wires, they would not open the door to new high-current applications.[14]

The extraordinary enthusiasm that greeted the arrival of high T_cs was no doubt due in large measure to their perceived economic interest and industrial possibilities, but it was also because they seemed to open up new avenues in the field of superconductivity itself. On the theoretical side, in particular, no major advances had occurred for more than twenty years. The years since 1987 have proven disappointing, however, and dramatic breakthroughs have been noticeably absent. Nevertheless, one of the new superconductors, Sr_2RuO_4, strontium ruthenate, which seemed unremarkable at the time of its discovery in 1994, has turned out to be truly exceptional, and not because of a record T_c—it's only 1.5K. Little by little, experiments have shown that its behavior is totally different from that of all other superconductors, even the high T_c ceramics whose crystal structure is very similar. It appears that the pairing process in this compound has a completely new symmetry that arises from the angular dependence of the wave functions of the electrons, called p waves, and the requirements of the Pauli exclusion principle. Thus, the spins of the two paired electrons tend to point in the same direction (i.e., parallel to each other) and not antiparallel, as they are in every other superconductor. The only analogous situation is superfluidity in ^{3}He, which was discovered in the 1970s in the millikelvin temperature range and found to be totally different from superfluidity in ^{4}He. The effect occurs in ^{3}He from the formation of pairs of atoms that have the p wave symmetry and the aligned spins that are seen for the Cooper pairs of electrons in strontium ruthenate.

The title and subtitle of this book suggest an unbreakable link between the development of low-temperature techniques and that of superconductivity. Certainly, this was true for much of the twentieth century, but here at the dawn of the twenty-first the two activities seem to be advancing independently. On the one hand, superconductivity progresses toward increasingly higher critical temperatures with techniques that have become more and more routine: using a superconducting coil in a magnet might involve some cost issues, but the auxiliary refrigeration hardware would be an insignificant part of typical laboratory infrastructure. Clearly, it's the same for electronic applications. On the other hand, progress in reaching extremely low temperatures has been most impressive, and it's measured no longer in decimals but in powers of ten: in the last twenty years temperatures have dropped from the millikelvin range (10^{-3}K) to microkelvins (10^{-6}K) and soon nanokelvins (10^{-9}K).

What is gained in coldness is lost in the mass of matter that gets cold, of course. Using techniques in which classical cryogenics plays no role, one cools small clusters of atoms, maybe a hundred to a thousand, to such low temperatures that the atoms are almost immobile and can be observed individually. The conquest and exploitation of these unimaginably low temperatures has been one of the quickest paths to a Nobel Prize in recent years.

No matter what method is used, the only way to cool even a small number of atoms is to transfer their energy somewhere. For these extremely low temperatures a "bath" of photons from one or several lasers, with a frequency matched to the properties of the atoms, is the best recipient. The atoms are locked in magnetic traps and have no choice but to transfer their last bit of thermal agitation energy to omnipresent photons. If the cooled atoms have integer spin, such as certain alkalines like cesium and rubidium, they cannot escape "Bose-Einstein condensation," a state we learned about from Fritz London — first observed in 1996 and which has since become the basis of an important subfield of physics. Don't be fooled by the term *condensation*: these condensed atoms are far apart from one another. As in Cooper pairs, the condensation takes place not in position space but in momentum space. All the atoms are in the same quantum state of lowest energy and overlap one another in a quasi-macroscopic state that is the condensate, visible as a kind of spot with fuzzy contours. Bose-Einstein purists would say that the superfluid ^{4}He state and a fortiori superconducting Cooper pairs are only a simulation of the true Bose-Einstein condensation now observed with strongly interacting cold atoms. The 1938 controversy between London and Landau continues afresh.

In 1908 Kamerlingh Onnes wanted to liquefy helium, but he surely did not imagine what would follow. Can we dream about what these cold, cold atoms might one day reveal?

NOTES

PREFACE

1. "Introductory Lecture on Experimental Physics," in *Scientific Papers of J. C. Maxwell*, ed. W. D. Niven, 2 vols. (Cambridge, 1890).

1. THE LOGIC OF LOW TEMPERATURE

1. K. Mendelssohn, *The Quest for Absolute Zero*, 2d ed. (London: Taylor and Francis, 1977), 10.

2. M. and B. Ruhemann, *Low Temperature Physics* (Cambridge: Cambridge University Press, 1937), 5.

3. From the *Memoires de chimie* of Lavoisier in *Oeuvres de Lavoisier*, 6 vols., vols. 1–4, ed. J. B. Dumas, vols. 5–6, ed. E. Grimaux (1862–1893; rpt., New York: Johnson Reprint Corp., 1965), 804. The citation is from vol. 2 of the 1862 edition. Much of Lavoisier's work was not published before he was guillotined. His wife ensured that it would become known by publishing a collection attributed to him with limited distribution in 1805; it was not put up for sale, and only a few collectors have copies. The 1862 edition was widely distributed.

4. Much to our surprise, these arguments continue. The history of science section of the Royal Institution met in October 1988 to discuss the history of low-temperature physics in Britain. A lively debate ensued about the relative merits of Dewar and W. Hampson, who seems to have invented the compressor that Dewar used to liquefy hydrogen. See also Sven Ortoli and Jean Klein, *Histoire et légendes de la supraconduction* (Paris: Calmann-Levy, 1988), 22.

2. PERPETUAL MOTION?

1. Emilio Segrè, *From X-rays to Quarks* (New York: W. H. Freeman, 1980), 223.

2. Hendrik B. G. Casimir, *Haphazard Reality, Half a Century of Science* (New York: Harper and Row, 1983), 161. This book includes a dedication "to the memory of my three teachers, Paul Ehrenfest, Niels Bohr, Wolfgang Pauli." It would be difficult to put oneself in better hands.

3. Mendelssohn, *Quest*, 79.

4. Ibid.

5. In 1925 Mrs. G. L. De Haas-Lorentz published a calculation that passed unnoticed on the attenuation of a variable magnetic field in the interior of a superconductor; it was a precursor of London's electromagnetic theory. (See chap. 6, n. 6.)

6. H. Kamerlingh Onnes, *Comm. Leiden* 140b (1914): 9; and 140c (1914): 21. Ampère's concept of magnetism included permanent surface currents. These articles are in English, but many of the early articles in the communications of the Leiden Laboratory were in French.

7. Cited by M. J. Klein in *Paul Ehrenfest: The Making of a Theoretical Physicist* (Amsterdam: North-Holland, 1970), 1:214; reprinted in R. de Bruyn Ouboter, "Superconductivity: Discoveries during the Early Years of Low-Temperature Research at Leiden 1908–1914," *IEEE Transactions on Magnetics*, MAG-23 (1987): 355.

8. H. Kamerlingh Onnes, "Nobel Lecture," <http://www.nobel.se/physics/laureates/1913/onnes-lecture.html>.

9. J. J. Thomson, *Phil. Mag.* 30 (1915): 192.

10. C. Benedicks, *Jahrb. d. Rad. u. Elekt.*, 13 (1916): 351; P. W. Bridgman, *Proceedings of the National Academy of Science* 3 (1916): 10; and *Phys. Rev.* 9 (1917): 269. See K. Gavroglu and Y. Goudaroulis, *Methodological Aspects of the Development of Low Temperature Physics* (Boston: Kluwer Academic Publishers, 1989), 70.

3. METALS AND THEORIES

1. Einstein said of Lorentz that at the turn of the century he was considered by theoretical physicists in every country as the leading mind among them and rightly so. *Mein Weltbild* (Zurich: Europa Verlag, 1953).

2. Segrè, *From X-Rays to Quarks*, 7.

3. In other words, Bloch considered electrons as simultaneously both propagating waves and particles, as Louis de Broglie had proposed. It may be relevant to note that Erwin Schrödinger was in Zurich at the time, that Peter Debye suggested that he present a seminar on de Broglie's thesis, and that the seminar was given on November 23, 1925, with Felix Bloch in the audience. See F. Bloch, *Phys. Today* 29 (December 1976): 23; and W. Moore, *Schrödinger: Life and Thought* (Cambridge: Cambridge University Press, 1989), 191.

4. The term employed in French at this time to designate the phenomenon that has always been called *superconductivity* in English was not fixed. Here Kamerlingh Onnes speaks (in French) of *superconductivité*; later, Edmond Bauer used *supraconduction* in his translation of Fritz London. It was only after the war that *supraconductivité* became established in French, like the German *Supraleitung*. For *superfluidity*, on the other hand, this term in English and the corresponding *superfluidité* in French were always used, no doubt because Kapitza, one of the discoverers, had used it in his first paper on the subject.

5. Lillian Hoddeson, Gordon Baym, and Michael Eckert, "The Development of the Quantum Mechanical Electron Theory of Metals: 1928–1933," *Rev. Mod. Phys.* 59 (1987): 1.

4. EXPERIMENTS AND THEIR INTERPRETATIONS

1. K. Mendelssohn and F. Simon, *Zeit. für Phys. und Chem.* B16 (1932): 72.

2. D. J. Quin III and W. B. Ittner III, *J. Appl. Phys.* 33 (1962): 748.

3. H. Kamerlingh Onnes, *Comm. Leiden Suppl.* 50a (1924): 3; W. Tuyn, *Comm. Leiden* 198 (1929): 3.

4. F. London, *Une Conception nouvelle de la supraconductibilité* (Paris: Hermann, 1937), 9.

5. Ibid., 7.

5. THE TRUE IMAGE OF SUPERCONDUCTIVITY

1. W. J. de Haas and H. Bremmer, *Comm. Leiden* 214d (1931): 37; and 220b (1932): 5.

2. E. Bauer, *Jour. de Phys. et Rad.* 10 (1929): 345.

3. W. Meissner and R. Ochsenfeld, *Naturwissenschaften* 21 (1933): 787.

4. C. J. Gorter and H. B. G. Casimir, *Physica* 1 (1934): 306.

5. C. J. Gorter, *Rev. Mod. Phys.* 36 (1964): 5.

6. M. and B. Ruhemann, *Low Temperature Physics*, 286.

7. W. Meissner and R. Ochsenfeld, *Zeit. für Tech. Phys.* 15 (1934): 507.

8. Y. N. Riabinin and L. V. Shubnikov, *Phys. Zeit. Sov.* 6 (1934): 557.

9. K. Mendelssohn and J. D. Babbitt, *Proc. Roy. Soc. London* A151 (1935): 316.

10. A. H. Wilson, *Reports on Prog. in Phys.* 3 (1936): 262.

11. Mendelssohn, *Quest*.

12. John Bardeen discusses this meeting in *Proc. Roy. Soc. London* A371 (1980): 77.

13. J. C. Slater, *Phys. Rev.* 51 (1937): 195; and *Phys. Rev.* 52 (1937): 214.

6. FRITZ

1. K. Gavroglu, *Fritz London: A Scientific Biography* (Cambridge: Cambridge University Press, 1995).

2. W. Heitler and F. London, *Zeit. für Phys.* 44 (1927): 455.

3. L. W. Nordheim, preface to the reprint edition of *Superfluids*, vol. 1: *Macroscopic Theory of Superconductivity*, by F. London (1950; rpt., New York: Dover, 1961), 6. The same text appears in *Phys. Today* 7 (July 1954): 16.

4. Simon revolutionized cryogenics with his invention of small helium liquefiers that were connected directly to the neck of the cryostat in which the low-temperature measurements were made. This type of instrument, just the opposite of the *big science* equipment in Leiden, later made it possible for laboratories without much experience in instrumentation to make their mark in low-temperature physics. This was the case in Oxford, where Simon took refuge in 1934. Just before the war Simon and Kurti carried out the first liquefaction of helium in France, at Meudon-Bellevue; they brought one of the liquefiers with them in their baggage.

5. R. Becker, G. Heller, and G. Sauter, *Zeit. für Phys.* 85 (1933): 772.

6. In his 1964 article in *Rev. Mod. Phys.* Gorter noted that G. L. de Haas-Lorentz (the daughter of Lorentz) had published an article on this subject in *Physica* in 1925 that was apparently completely overlooked. She considered the question posed by her husband of whether a variation in the strength of an external magnetic field could affect a superconductor and whether any effect could be completely canceled by a surface current, as Kamerlingh Onnes thought. She obtained the same expression for what came to be called the penetration depth as London obtained ten years later.

7. Moore, *Schrödinger*, 265.

8. Concerning F. A. Lindemann, see ibid., 267. Born of an American mother and an Alsatian father, he was all his life an intimate friend and, as shown by C. P. Snow, a deplorable scientific advisor to Winston Churchill. See C. P. Snow, *Science and Government* (Cambridge: Harvard University Press, 1961).

9. Moore, *Schrödinger*, 269.

10. The refugee cryogenicists had much more influence on the Clarendon Laboratory at Oxford than might be apparent: Ralph Clifton, the director of the laboratory for forty years until 1915, was known for his resolute opposition to all experimental research activity. Lindemann took over in 1919.

11. Given the economic situation in Britain, the minister of labor was initially opposed to the idea of accepting a large number of German scientists.

12. Edith London in *Superfluids*, 1:xii–xiii.

7. NOT YOUR EVERYDAY LIQUID

1. Mendelssohn, *Quest*, 89.

2. W. H. Keesom and K. Clusius, *Comm. Leiden* 219e (1932): 42.

3. H. Fröhlich, *Physica* 4 (1937): 639.

4. London considered it highly unphysical that N atoms would preferentially occupy one of the two subnetworks. See F. London, "The Lambda-Point Phenomenon of Liquid Helium and the Bose-Einstein Degeneracy," *Nature* 141 (1938): 643.

5. J. C. McLennan, H. D. Smith, and J. O. Wilhelm, "The Scattering of Light by Liquid Helium," *Phil. Mag.*, ser. 7, 14 (1932): 161.

6. In his famous account of the first liquefaction of helium, Kamerlingh Onnes reported that, as he tried in vain to solidify the helium by lowering the temperature down to about 1K, his colleague, J. P. Kuenen, remarked to observers that the meniscus looked nothing at all like the meniscus of liquid air or liquid hydrogen. The significance of this first observation that liquid helium was unlike other liquids was not recognized.

7. The notion that superconducting electrons follow linear channels, as suggested by J. Frenkel (*Phys. Rev.* 43 [1933]: 908), is related to such considerations. The idea of a geometric order in ordinary space is strongly fixed in our intuition; even specialists in superconductivity would call it common sense.

8. THE RUSSIAN COLD

1. Cited by F. Kedrov, *Six physiciens à la découverte de l'atome*, French trans. (Moscow: Ed. MIR, 1979), 113.

2. The memories of Olga Trapeznikova are included in *L. V. Shubnikov: Collected Works and Memoirs* (Kiev: Physical-Technical Institute of Low Temperatures of the Ukrainian Academy of Sciences / Naukova Dumka, 1990), 256 (in Russian).

3. Tatiana Afanassjeva was also trained in physics. Ehrenfest met her in Vienna and married her in 1904 while working on his thesis with Boltzmann.

4. Trapeznikova, *Shubnikov*, 273. The theorist I. E. Tamm also visited Leiden that year, before Trapeznikova and Shubnikov set off on vacation for Paris and Brittany.

5. A. Livanova, *L. D. Landau* (Moscow: Ed. MIR, 1978).

6. Landau introduced the notion of what is called the density matrix in quantum mechanics at the same time as Felix Bloch. This information is taken from a detached assessment of Lev Landau by F. Janouch, in *Lev Landau: His Life and Work*, CERN Publication 79–03 (Geneva: CERN, 1979).

7. L. D. Landau and L. Rosenkevitch, *Zeit. für Phys.* 78 (1932): 847.

8. Janouch, *Landau*, 3.

9. The seven volumes of Landau's treatise were written in close collaboration with E. Lifschitz, who modestly considered himself to be Landau's scribe. It has been continuously reprinted and brought up to date, and it still represents a summary of knowledge indispensable for any theorist.

10. Trapeznikova (*Shubnikov*, 290) says Weissberg arrived in Kharkov in 1932.

11. Ruhemann's father was the chaired professor of organic chemistry at Cambridge. According to his son, he was nevertheless subject to numerous demonstrations of xenophobia during World War I, which led him to return to his native country at the end of the war.

12. Trapeznikova, *Shubnikov*, 280.

13. A. Weissberg, *The Accused* (New York: Simon and Schuster, 1951), 299.

14. After three years in prison, Weissberg was handed over to the Gestapo with the signing of the Soviet-German treaty. He was imprisoned and tortured in Warsaw but managed to escape and joined the Polish resistance, with which he fought until the German surrender. He reached Sweden and England and, in 1948, again met up with Koestler and Ruhemann. Koestler had already written *Darkness at Noon*; Ruhemann maintained some illusions about the Soviet Union.

9. IN CAMBRIDGE IN SPITE OF STALIN, IN MOSCOW BECAUSE OF STALIN

1. Bristol was very active in solid-state physics; Nevill Mott began to teach there in 1935.

2. The roots of cryogenics in the Soviet Union and in Oxford thus lay in Germany.

3. G. Gamow, *My World Line: An Informal Autobiography* (New York: Viking Press, 1970), cited by G. M. Spruch, *Phys. Today* 32 (September 1979): 39. Gamow, a scientist with an overwhelming imagination, also wrote remarkable popular science books (the *Mr. Tompkins* series) that made him known to the general public. Gamow arrived at the University of Leningrad at the same time as Landau and knew Trapeznikova there (*Shubnikov*, 259).

4. Private conversation, London, October 26, 1988.

5. This had already happened to the famous physicist Lazare Carnot and many others during and after the French Revolution.

6. Letter of May 7, 1935, in *Kapitza in Cambridge and Moscow*, ed. J. W. Boag, P. E. Rubinin and D. Shoenberg (Amsterdam: North-Holland, 1990), 319.

7. Letter of May 14, 1935, in *Kapitza*, 323.

8. Private conversation, London, October 26, 1988.

9. Will Brownell and Richard N. Billings, *So Close to Greatness: A Biography of William C. Bullitt* (New York: Macmillan, 1988). See also Ronald Steel, "The Strange Case of William Bullitt," *New York Review of Books* 35, September 14, 1988, 18.

10. P. A. M. Dirac is the well-known theorist. E. D. Adrian won the Nobel Prize in 1932. Later he became rector of Cambridge and, in the 1960s, president of the Royal Society.

11. D. Shoenberg, private discussion in Brighton during the Nineteenth International Conference on Low Temperature Physics, August 16–22, 1990.

12. Shoenberg was the first to determine a Fermi surface, the representation of the motion of electrons in a metal that has proved so useful. Shoenberg's result was based on his experiments in Cambridge in 1940 related to the de Haas–van Alphen effect and was thus the fruit of a long line of work by many experimenters, starting in Leningrad (with bismuth crystals), continuing in Leiden and Kharkov (by Shubnikov), and finally completed in Moscow and Cambridge.

13. G. G. Efimovitch, "In the dossier of the KGB concerning the academician L. D. Landau," *Voprosy Istorii* 8 (1993): 112. See also an article by the same author in an edition of *Izvestia* the same year, "At Whose Side Did Lev Landau Sit?" (*Za chto cidel Lev Landau*).

14. Concerning Fiodor Raskolnikoff, see I. Deutscher, *Stalin: A Political Biography*, 2d ed. (Oxford: Oxford University Press, 1967), 370.

15. Fock himself had been arrested the previous year, and Kapitza had protested to a leader of the education ministry. Kapitza had also written to Molotov in 1936 to complain about the virulent press campaign launched by *Pravda* against N. N. Luzin, a famous mathematician.

16. *Kapitza*, 348.

17. See Ref. 11.

18. P. Wright and P. Greenglass, *Spycatcher* (Richmond, Victoria, Aus.: Heineman, 1987), 259.

19. In fact, according to Vitali Ginzburg, Kapitza did not prepare the experiment himself, but, when it was all ready, he removed the physicist who had set it up (private conversation, September 25, 1991, in Paris).

20. *Kapitza*, 348. For example, Tupolev, the builder of airplanes, and his team continued to design airplanes, but they had no paper—they made do with wooden planks and carpenters' pencils.

21. Ibid., 350. Even though Kapitza defended only the most well-known scientists, he always did it with a remarkable sense of timing. Learning just after the German invasion of France that Paul Langevin had been imprisoned, he wrote immediately to Molotov: "Moscow, November 10, 1940. I just received a telegram from Cambridge (from Dirac) who said that Professor P. Langevin is in prison in Paris. Langevin is a great scientist and a great friend of the Soviet Union. I like him very much and he is an honest man. I would like our friends to know that we appreciate their friendship and I would hope that something can be done for him." Left implied was the reason—the German-Soviet pact. Around the same time, the secretary of the French Academy of Sciences was urged to intervene on behalf of Georges Bruhat, a talented physicist who was also imprisoned. According to Jean-Paul Mathieu, the secretary replied that he had an illness that kept him confined to his bed.

22. Someone in Landau's professional entourage reported all his private comments about politics in general and about nuclear research in particular to the KGB throughout his lifetime.

10. SUPERFLUIDITY: THEORIES AND POLEMICS

1. Mendelssohn, *Quest*, 246.

2. T. S. Kuhn, *The Structure of Scientific Revolutions* (Chicago: University of Chicago Press, 1962).

3. F. London, *Superfluids*, vol. 2: *Macroscopic Theory of Superfluid Helium* (New York: John Wiley and Sons, 1954), 17. This citation is emphasized by K. Gavroglu (National Technical University of Athens) and Y. Goudaroulis (Aristotle University of Thessalica) in *Ann. Science* 43 (1986): 137. These two authors have written several articles on the history of superfluidity and its epistemological implications. See also *Ann. Science* 45 (1988): 367; and an article in *Janus* in 1989, "From Physica to Nature: The Tale of a Most Peculiar Phenomenon."

4. A. Einstein, "*Quantumtheorie des einatomigen idealen Gases,*" *Sitzungberichte der Preussichen Akademie der wissenschaften Phys-Mat Klasse* (1925): 3. Bose statistics applies to particles with integer spin, including zero, whereas Fermi-Dirac statistics applies to particles with half-integer spin such as electrons.

5. Cited by D. Pines, *Phys. Today* 34 (November 1981): 119.

6. In a note to the French Academy of Sciences at a time when there was not one helium liquefier in France: L. Tisza, "Concerning Thermal Superconductivity," *Comptes Rendu de l'Academie des Sciences* 207 (1938): 1035. Even the title of this article shows that the analogy was something physicists often thought about.

7. The *Journal of Experimental and Theoretical Physics*, which was first published in 1955, is the English translation of the corresponding journal in Russian.

8. L. D. Landau, *Phys. Rev.* 75 (1949): 884. There is absolutely no doubt that Landau had been thinking the same thing since 1941. Note that Landau was authorized to publish this review article on helium in an American journal after the war. On the other hand, his works on superconductivity, published in Soviet journals, were essentially unknown in the West (see chap. 11).

9. L. D. Landau, *Zh. Eksp. Teor. Fiz.* 5 (1941): 71; V. L. Ginzburg, *Zh. Eksp. Teor. Fiz.* 7 (1943): 305; V. Peshkov, *Zh. Eksp. Teor. Fiz.* 10 (1946): 389; F. London, *Proceedings of the International Conference on Fundamental Particles and Low Temperatures*, Cambridge, U.K., July 22–27, 1946 (New York: American Physical Society, 1947). The Soviets did not participate in this conference. L. Tisza, *Proceedings of the International Conference on the Physics of Low Temperature*, September 6–10, 1949. These citations have been collected in K. Gavroglu and Y. Goudaroulis, *Ann. Science* 45 (1988): 367.

10. N. Bogoliubov, *Zh. Eksp. Teor. Fiz.* 11 (1947): 23.

11. Ginzburg was not permitted to visit the "Installation," the name given to the group of laboratories that carried out bomb research. Vitali Ginzburg, private conversation in Paris, September 25, 1991.

12. Andrei Sakharov, *Memoirs*, trans. Richard Lourie (New York: Knopf, 1990), 131.

13. Ibid.

14. We beg the reader's indulgence for our strong insistence on the fact that these names strictly have no meaning outside the context in which they were first used. There are just too many examples of words transplanted from one field to another with no assurance of retaining the old meaning in the new context. It is true, however, that, inspired more by the media than by the exchange of ideas, physicists themselves are happy to use graphic words or acronyms such as *chaos, catastrophe, charm, frustration*, and *neural networks*.

15. The spectrum of elementary excitations determined by low-energy neutron diffraction allows a check on the validity of Landau's model, because it can be analyzed to determine the flow velocity at which helium stops being superfluid. Unfortunately, the predicted value (about 60 m/s) is about 10,000 times the measured one. The American physicists Richard Feynman and Lars Onsager suggested a possible explanation of this discrepancy in the early 1950s, based on the collective vortex motion of all the atoms in a container. These vortices, actual macroscopic quantum states, were observed only in the 1960s. They represent one of the bridges that link superfluidity and superconductivity. Although they were not predicted by Landau, they do not contradict his theory.

11. THE WAR, THE BOMB, AND THE COLD

1. Casimir, *Haphazard Reality*, 202.

2. Vitali Ginzburg, private conversation in Paris, September 16, 1991.

3. Von Laue, known above all for his work on crystallography, published all his articles on superconductivity in 1942 to defy the Nazis' prohibition on references to "Jewish" science. Von

Laue even intervened with a colleague so that Fritz Houtermans could work again in government laboratories. Houtermans was a nuclear physicist who had taken refuge in the Soviet Union in 1933 and had been turned over to the Germans at the same time as Alexander Weissberg, right after the signing of the Russian-German pact. Concerning Houtermans, see D. Irving, *La Maison des virus* (Paris: Robert Laffont, 1968), 97, 109; Casimir, *Haphazard Reality*, 133, 220; and Richard Rhodes, *The Making of the Atomic Bomb* (New York: Simon and Schuster, 1986), 370. Houtermans had worked with Shubnikov and I. V. Kurchatov, the father of the Soviet atomic bomb program. The group published an article written in German in the *Soviet Journal of Physics* on the absorption of thermal neutrons in silver at low temperatures: V. Fomin, F. G. Houtermans, I. V. Kurchatov, and L. V. Shubnikov, *Phys. Zeit. Sov.* 10, 103. At the end of the war von Laue was imprisoned in England with the physicists responsible for the German nuclear program — W. Heisenberg, O. Hahn, W. Gerlach, and C. F. von Weisäcker. He was the only German physicist for whom Einstein maintained his respect.

4. E. Justi, *Zeit. für Phys.* 44 (1943): 469.

5. Irving, *Maison*, 164.

6. Rhodes, *Bomb*, 343.

7. Irving, *Maison*, 100; Rhodes, *Bomb*, 340.

8. Biography of Heinz London (1907–1970) by W. M. Fairbank and C. W. F. Everett in *Dictionary of Scientific Biographies*, ed. Charles Coulston Gillespie (New York: Charles Scribners and Sons, 1975).

9. After the war Samuel A. Goudsmit became the editor of the *Physical Review*. In 1947 he published a lively, frank, and controversial account of his mission, *Alsos* (New York: H. Schuman, 1947). A second edition includes a new foreword by R. V. Jones (Los Angeles: Tomash Publishers, 1983). It triggered considerable debate about the role of physicists in the Nazi program and the value of their work.

10. Irving, *Maison*, 247.

11. Goudsmit, *Alsos*, 56. In contrast to uranium and plutonium, thorium is not fissile; its nucleus cannot be split by slow neutron bombardment. The atom of thorium with mass number 232 can, however, absorb a low-energy neutron, which, after emitting two successive electrons, leads to protoactinium and then to uranium 233, which is fissile.

12. Goudsmit (*Alsos*, 65) presents this story as if the toothpaste were a new product, but this seems to be incorrect. Rhodes (*Bomb*, 283) indicates that Doramad existed before the war.

13. Rudolf Peierls, the first physicist on the Anglo-Saxon side to estimate the critical mass of ^{235}U necessary for a chain reaction, never shared this pessimistic view of the German atomic bomb effort. He became convinced early on that the bomb effort was not a priority for the Germans, even though the German physicists were convinced they would be first, and few had any moral scruples about it. See *Operation Epsilon: The Farm Hall Transcripts* (Bristol: British Institute of Physics, 1993). Peierls noted that the list of physics courses taught in Germany published every semester in the *Physikalische Zeitschrift* revealed that almost all physicists were fulfilling their teaching obligations and teaching their usual courses. A similar survey in the United States or England would have given rather different results. See Peierls, "The Bomb That Never Was," *New York Review of Books*, April 22, 1993, 7.

14. Experimentalists' judgments about theorists' ideas can also be criticized. Around this time Fermi thought it wouldn't be feasible to separate the isotopes of uranium.

15. *Science and Academic Life in Transition: E. Piore*, ed. Eli Ginzburg (New Brunswick, N.J.: Transaction Publishers, 1990), 12.

12. RADAR AND SUPERCONDUCTIVITY

1. C. P. Snow, an advisor for scientific affairs to the British government and the person responsible for the idea of two opposing cultures, science and the humanities, during the 1960s, viewed the magnetron as the "most important British scientific contribution during the war against Hitler" (Snow, *The Physicists* [London: Little, Brown, 1981], 105).

2. McLennan had already made measurements at 24 gigahertz (2.4×10^{10} hertz) in 1930–1931, but, according to Fritz London, they were inconclusive because the surfaces of his sample were not clean enough. See J. C. McLennan, A. C. Burton, A. Pitt, and J. O. Wilhelm, *Proc. Roy. Soc. London* 136 (1931): 52; and London, *Superfluids*, 1:90.

3. Casimir's experiments were based on measuring the change in the mutual inductance of two coils wound around the same cylindrical superconductor (H. B. G. Casimir, *Physica* 7 [1940]: 887). The mutual inductance measures how a change in magnetic field in one coil affects the current in the other; it depends sensitively on the size of the coils. Much later E. Laurmann and D. Shoenberg discovered an important reason why Casimir's experiments were inconclusive (E. Laurmann and D. Shoenberg, *Proc. Roy. Soc. London* A198 [1949]: 560). Reducing the pressure of the helium bath to lower the temperature causes a slight stretching of the coils, which tends to cancel the effect of the changing penetration of the magnetic field that Casimir was trying to measure. The cleanliness of the surface of the superconductor and its crystalline orientation are also important (E. Lauerman and D. Shoenberg, *Nature* 160 [1947]: 747).

4. London, *Superfluids*, 1:41.

5. If the beads are large, the expulsion of the magnetic field in each one gives a signal proportional to its volume (R^3), whereas the signal is proportional to R^5 if the beads are small compared to the penetration depth. The signals themselves are small, with large variations from one bead to the next — Shoenberg's experiments, with the techniques and equipment available in 1940, were not easy. In the 1960s the superconductivity group at Orsay repeated the experiment, this time to show that type I superconductors could under certain conditions stay superconducting even above the critical magnetic field, the way *superheated* water can sometimes reach temperatures above the boiling point. Making the samples with tiny drops of mercury was difficult, until someone, no doubt a reader of Casanova or Guy de Maupassant, remembered that mercury salts in the form of a cream were once the preferred treatment for syphilis. (The mercury in these salts was responsible for many of the associated problems of the nervous system.) Orsay was still at the time a rural hamlet, oriented more toward the charms of the nearby Valley of the Chevreuse than the more suspect charms of technology parks. A short trip to the local pharmacy uncovered one of the last samples of the venerable cream, a little bead of which, inserted into a coil and dipped in helium, allowed the first measurements of the limits on the "superheated" superconductivity, which had seemed beyond experimental reach.

6. D. Shoenberg, *Proc. Roy. Soc. London* A175 (1940): 49.

7. I. K. S. Kikoin and S. W. Gubar, *J. Phys. USSR* 3 (1940): 33, cited by London, *Superfluids*, 1:83n.

8. A. B. Pippard, *Proc. Roy. Soc. London* A191 (1947): 370, 385, 399.

9. Shoenberg, meanwhile, in collaboration with M. Désirant, tried to make a small sample of mercury seem like a massive superconductor. He measured the variation of the magnetic penetration depth as a function of the temperature of a capillary filled with mercury. He reduced the volume of the mercury column to a hundredth of a millimeter, but the measured variation was still only 1 percent of the radius of the capillary. See M. Désirant and D. Shoenberg, *Nature* 159 (1947): 201; *Proc. Roy. Soc. London* 60 (1948): 413.

10. F. London, *Superfluids*, 1:90.

11. G. E. H. Reuter and E. H. Sondheimer, *Nature* 161 (1948): 394; *Proc. Roy. Soc. London* A195 (1948): 336.

12. Once again it may be useful to point out that ordinary language can be misleading. The Heisenberg inequalities, routinely called the "Heisenberg uncertainty relations," are commonly misinterpreted by the public to mean that the results physicists achieve in the quantum world are always uncertain. Likewise, the "abnormal skin" has nothing to do with a pathological behavior of a metal. On the contrary, it's the extreme purity of the metal and the lack of any faults in the crystal that allow the abnormal skin.

13. London's formula suggested that λ was inversely proportional to the density of superconducting electrons, so that λ would increase (and the magnetic field would thus penetrate further) as the density decreased.

14. Pippard, *Proc. Roy. Soc. London* A216 (1953): 547.

15. Ibid. The inverse of the coherence range is equal to the sum of the inverses of two lengths: the coherence range of the absolutely pure superconducting metal and the mean free path of the electrons.

16. Strictly speaking, this argument is correct only for superconductors that, like mercury, have an abrupt transition to the normal state when they are in an increasing magnetic field. Pippard did his experiments with such metals, which are now called type I superconductors. It is questionable whether he would have succeeded in these experiments that proved so important if he had used type II superconductors, in which, as we shall see, the coherence length is shorter than the penetration depth.

17. H. Fröhlich, "Theory of the Superconductive State," *Proceedings of the International Conference on Low Temperature Physics*, Oxford, August 22–28, 1951, ed. D. F. Brewster (1951; rpt., Eindhoven: North-Holland, 1990), 114.

18. London, discussion after the article by Fröhlich (ibid., 110).

19. This choice of epithet clearly separates solid-state physicists, so attached to the nitty-gritty of their materials, from their colleagues in particle physics, who emphasize the more abstract and philosophical aspects of their search, with particles labeled *charm*, *beauty*, and *truth*.

20. Pippard, *Proc. Roy. Soc. London* A216 (1953): 547.

21. Awareness of this important result did not spread quickly; even the eminent B. Serin still believed in the T^3 dependence in 1955. (See chap. 13, n. 15.)

13. THE IONS ALSO MOVE

1. L. W. Alvarez and R. Cornog, *Phys. Rev.* 56 (1939): 379; and *Phys. Rev.* 56 (1939): 613. One part in ten million of terrestrial helium is ^{3}He. H. Bethe and R. Cornog report on their recollections of this period in *Discovering Alvarez*, ed. W. Peter Trower (Chicago: University of Chicago Press, 1987), 25.

2. Fritz London again, but that's not surprising since this was at the core of his preoccupations, and his arguments with the school of Landau and with Tisza were at their peak (London, *Nature* 163 [1949]: 694).

3. Both London (F. London and O. K. Rice, *Phys. Rev.* 73 [1948]: 1188) and Tisza (*Phys. Today* 1 [August 1948]: 4) predicted that ^{3}He could not even exist as a liquid at atmospheric pressure because of its large zero-point energy. As low-temperature techniques improved, however, not only was liquid ^{3}He discovered, but in 1971 it was also found to be a superfluid below 0.025K. Later several superfluid phases were found, but that's another story.

4. Fröhlich, *Proc. Roy. Soc. London* A371 (1980): 102. Harwell is a center for nuclear research in Britain.

5. The references to the first two articles on the isotope effect are: E. Maxwell, *Phys. Rev.* 78 (1950): 477; and C. A. Reynolds, B. Serin, W. H. Wright, and L. B. Nesbitt, *Phys. Rev.* 78 (1950): 487.

6. During the 1951 Oxford conference mentioned in the last chapter, which took place just after the first measurements of the isotope effect, Pippard made an astute remark that turned out to be the first glimmer of his notion of the coherence range. He observed that even for elements such as tin, with a large number of isotopes, the width in temperature for the normal-to-superconducting transition was very narrow. Superconductivity thus implies that a great number of atoms must be involved simultaneously over distances that he estimated as larger than 10^{-4} cm (Brewster, *Oxford Proceedings*, 118).

7. Serin died of a heart attack in England in the 1970s while running for a train that had just left. The building that houses the Physics Department at Rutgers University is named for him.

8. J. Bardeen, *Proc. Roy. Soc. London* A371 (1980): 77. This is a contribution by Bardeen to the history of physics entitled "Reminiscences of the Early Days of Solid State Physics." The citation is on p. 82.

9. Ibid. The work of Slater to which Bardeen refers had been published in two successive articles: *Phys. Rev.* 51 (1937): 195; and *Phys. Rev.* 52 (1937): 214.

10. Bardeen, *Phys. Rev.* 59 (1941): 928.

11. Bardeen, *Phys. Rev.* 79 (1950): 167; Fröhlich, *Phys. Rev.* 79 (1950): 845.

12. B. Serin, *Prog. in Low Temp. Physics* 1 (1955): 139.

13. Ibid., 143.

14. Ibid.

15. Ibid., 144. In addition, in the same article, Serin considers the difference in specific heat between two phases of a normal metal. He includes only one term in T^3—that is, he takes into account only the effect of the phonons, as if the electrons did not contribute to the specific heat. This was based on the old result of J. A. Kok (*Physica* 1 [1934]: 1103), which indicates that the specific heat in the superconducting phase varies as T^3. The measurements that supported this finding, however, were not sufficiently sensitive. Because of the energy gap, Kok's expression is not valid and, in principle, should not be used this way. The correct theoretical expression would only come later with the work of Bardeen, stimulated by the isotope effect.

14. EAST IS EAST, AND WEST IS WEST

1. John Bardeen paid homage on several occasions to the work of Fritz London, who was for him a source of permanent inspiration.

2. V. L. Ginzburg and L. D. Landau, *Zh. Eksp. Teor. Fiz.* 20 (1950): 1064; reprinted in L. D. Landau, *Collected Papers*, ed. D. Ter Haar (New York: Gordon and Breach, 1965), 546.

3. P. W. Anderson, in *Superconductivity*, ed. R. D. Parks, vol. 2 (New York: Marcel Dekker, 1969), 1347.

4. Ginzburg and Landau, *Zh. Eksp. Teor. Fiz.* 20 (1950): 1064.

5. The order parameter, introduced in chapter 3, is supposed to provide a quantitative measure of the relative amount of ordered (superconducting) phase and disordered (normal) phase. Its choice is somewhat arbitrary and intuitive. Because a metal can be in a superconducting state

without any actual current—for example, if a loop of niobium is cooled below T_c with no external magnetic field present—the density of supercurrents cannot be a good order parameter. This, however, was Landau's original choice. London's work had since established the idea of an electronically ordered state characteristic of the superconducting state and responsible for the Meissner effect. We don't know for sure, but it is reasonable to think that Landau now had in mind London's proposals about what a coherent explanation of superconductivity should involve.

6. John Robert Schrieffer is generally known as Robert or Bob; he doesn't use his first name.

7. J. R. Schrieffer, *Theory of Superconductivity*, 3d ed. (Redwood City, Calif.: Addison-Wesley, 1983), 19; emphasis added. Algebraic expressions have been omitted.

8. If we go back to the bowling metaphor, we can imagine that, instead of the usual ten pins set up in one lane for one bowler, the bowler commands an entire huge surface with thousands of pins set up in many rows and columns. The order parameter at a point might then be the percentage of pins left standing in a circle of 1 m radius about that point. The order parameter would then be unity everywhere before the bowler started, and it would decrease to zero as pins fell. It would not be the same everywhere until all the pins were down, however, and, with a bowler who liked to face left, might be close to zero for most points on his left while it remained close to unity for most points on his right.

9. The use of complex numbers is an efficient way to manipulate quantities that depend on two distinct numbers, such as the position of a point on the earth in terms of latitude and longitude. It's particularly useful for quantities that vary periodically in space or time, like the face of the moon—where we explicitly talk about its phase—or the height of the sea. To describe the state of the tide at a specific point and time, two numbers are necessary: the total *amplitude* of the tide (i.e., the difference in the height of the sea between low and high tide that day) and the *phase*, which is the difference in time between the last low tide and the actual time. The combination of amplitude and phase constitutes a complex number.

10. A. A. Abrikosov, *Zh. Eksp. Teor. Fiz.* 32 (1957): 1442; English trans., *JETP* 5 (1957): 1174.

11. Jean Matricon, *Cuisine et molécules* (Paris: Ed. Echo Hachette and Palais de la Découverte, 1990).

12. The paradoxical fact that a magnetic field can have an effect in a region where the field is zero is not something specific to superconductors; it was discovered by Faraday and described mathematically by Maxwell. If a current flows in a coil tightly wound in the shape of a long cylinder, which physicists call a *solenoid*, the magnetic field inside the coil is large, but outside it is essentially zero. Now consider a large loop of wire encircling the solenoid close to its midpoint: the entire loop is in the region where the field due to the solenoid is zero. Even so, if the magnetic field inside the solenoid is changing, a current will flow in the loop; the current will not flow if the field is constant. To describe this interesting example of *induction*, Maxwell introduced a new quantity called the *vector potential*, which must be nonzero in a region where the magnetic field is zero. Apparently a purely mathematical construction for Maxwell, the value of the vector potential cannot be measured in any experiment. Nevertheless, particularly in the formalism of quantum mechanics, it has become a standard physical quantity. Far from the center of a vortex, in a region where the magnetic field and its associated supercurrents are zero, the vector potential is not zero.

13. R. P. Feynman, "Atomic Theory of the Two Fluid Model of Liquid Helium," *Phys. Rev.* 94 (1954): 262.

14. London, *Superfluids*.

15. W. Heisenberg, "The Electron Theory of Superconductivity," seminar paper presented at Cambridge in 1948, in W. Heisenberg, *Two Lectures* (Cambridge: Cambridge University Press, 1949). Heisenberg was trying to establish a parallel between ferromagnetism and superconductiv-

ity. The idea of a crystal of electrons might be able to explain infinite conductivity: if the electron crystal had the same configuration as the ionic lattice, the two might slide in between each other without collisions, as the teeth of one comb might slide through the teeth of another. While the geometry seems satisfactory, this idea cannot explain the Meissner effect. Curiously enough, however, this model, which gives a completely false idea of the phenomenon, was the only model presented in a French television program on superconductivity, and the program won the prize for the best scientific program in 1989.

16. Readers will no doubt be surprised by Heisenberg's blindness (especially if they realize that Meissner, who lived to be very old, was once more involved in research). Nevertheless, it is important to keep in mind that physicists at the time still did not know where the field of superconductivity was headed. Even Bardeen, for whom the perspective opened by London was *the* source of inspiration, considered that the "theory of London had not received a good experimental confirmation" (Bardeen, in *Handbuch der Physik*, vol. 15, ed. S. Flügge [Berlin: Springer Verlag, 1956], 275, 284). This was just before he did his decisive work with Cooper and Schrieffer. Bardeen recognized that obtaining the right order of magnitude for the penetration depth was a success of this theory but noted the contradiction with the experiments that demonstrate that the penetration depth depends on the orientation of the metallic crystal, as Pippard had shown for tin (Pippard, *Proc. Roy. Soc. London* 203 [1950]: 98). This was, however, a secondary issue, since von Laue had demonstrated that London's theory could be generalized to anisotropic superconductors — that is, those whose properties depend on the direction relative to the axis of the crystal (M. von Laue, *Ann. Phys. Lpzg.* 3 [1948]: 31).

17. London, *Phys. Rev.* 74 (1948): 562. On another occasion London would remark that models that take perfect conductivity as the fundamental characteristic of a superconductor must admit that there are an infinite number of possible models of the current. In his terms this would "offend the good taste of most physicists" (cited by B. S. Chandrasekhar in Parks, *Superconductivity*, 1:35).

18. London, *Superfluids*, 1:1–3.

19. This strict notion of rigidity was soon overtaken by further theoretical developments, which showed that macroscopic quantum states with spatial modulation were acceptable. Abrikosov's vortices in superconductors, ironically unknown in the West for some years, are one example.

20. A point in *ordinary three-dimensional space* can be defined with respect to three axes labeled $(x, y,$ and $z)$ oriented at 90° to one another, so the position of an electron in a metal can be specified this way. The velocity of an electron can be represented by an arrow (vector $\mathbf{v}$) that points in the direction of its motion; the length of the arrow is proportional to the speed of the electron. The tip of the arrow is a point in *velocity space*, which can be defined with respect to three axes labeled $(v_x, v_y,$ and $v_z)$. The maximum value of the velocity of an electron is limited, but the arrows can point in any direction. In an isotropic material (a material that has the same properties in every direction, such as a sphere) the velocity arrow is equally likely to point in any direction, so, because there are so many electrons, the tips of the arrows corresponding to the maximum speed will cover a spherical surface in velocity space. The Fermi surface for such a material is, then, a sphere. Often the Fermi surface has a more complicated shape, particularly for anisotropic materials (materials that have different properties in different directions, such as mica, in which layers slide easily along one another parallel to the surface but not perpendicular to it). The study of the exact form of the Fermi surface reveals considerable information about a solid. One final detail: instead of using velocity vectors, condensed matter physicists prefer to multiply the velocities by the mass m of the electrons. The product of m and the vector $\mathbf{v}$ defines the momentum vector, and then they speak of momentum space rather than velocity space. The shape of the Fermi surface is unchanged.

21. Parks, *Superconductivity*, 1:4.

22. To complete the story about this model of Heisenberg, we should note that it was developed by H. Koppe, *Ann. Physik* (6) 1 (1947): 405; and *Zeit. für Naturforschung* 3a (1948): 429. M. R. Schafroth is another theorist using the hydrodynamic approach who will appear later in this account; he was not one to mince words. In his last article, published postmortem, concerning the work of Heisenberg and Koppe, he remarked that "the foundations of this theory appear quite shaky" (Schafroth, *Solid State Phys.* 10 [1960]: 304). This model indeed predicted a gap in energy between the superconducting state and the normal state, but in 1950 this was the least that could be expected of any model proposed by theorists at the cutting edge of this subject.

15. NOW, HOW TO GRAB THE TIGER BY THE TAIL?

1. D. Millar, ed., *The Messel Era: The Story of the School of Physics and Its Science Foundation within the University of Sydney, 1952–1987* (Sydney, Aus.: Pergamon Press, 1987).

2. Schafroth, *Phys. Rev.* 96 (1954): 1149.

3. Schafroth, *Helv. Phys. Acta* 24 (1951): 645.

4. Schafroth, *Phys. Rev.* 100 (1955): 463.

5. Schrieffer, *Superconductivity*, 32.

6. The interaction between electromagnetic radiation and matter represents probably the most elegant application of Feynman diagrams. Feynman was awarded the Nobel Prize in 1965 for this theory, called quantum electrodynamics. In its original form it is extremely abstract, but Feynman wrote a version for the general public that we think is one of the most outstanding scientific popularizations ever written. It is called *QED: The Strange Theory of Light and Matter* (Princeton: Princeton University Press, 1988).

7. The quantity in helium vortices that is quantized—that is, which takes on a limited set of discrete values—is called the *vorticity*, a measure of the speed with which the liquid turns around the axis of the vortex. It is a macroscopic quantity, because the rotation of the fluid in a vortex extends right up to the walls of the helium container, but its values must be integer multiples of h/M, where h is Planck's constant and M is the mass of helium. The fact that the minimum value of the vorticity is larger than zero means that the flow of liquid helium at low velocity occurs without "friction"—that is, with strictly zero viscosity—which means superfluidity, discovered eighteen years earlier by Kapitza and, independently, by Allen and Misener.

8. Bardeen and Pines, *Phys. Rev.* 99 (1955): 1140.

9. Feynman, *Rev. Mod. Phys.* 29 (1957): 209.

10. Ibid., 212.

11. Anderson, in Parks, *Superconductivity*, 2:1343.

12. Schrieffer had plenty of time to think about the words of Feynman because he was asked to write them up for publication in the *Proceedings*. He first prepared a word-by-word copy of the tape recording, but Feynman was enraged because his words didn't make complete sentences. He told the editor he wanted nothing to do with this fellow Schrieffer and would forbid publication of the manuscript. Finally, Schrieffer went back to the typewriter and provided the new version that we know (private conversation with Schrieffer in Grenoble, July 7, 1994, during a conference on high T_c superconductivity).

16. JOHN BARDEEN'S RELENTESS PURSUIT

1. Bardeen, Walter Brattain, and William Shockley discovered the transistor effect at the end of December 1947. They were awarded the Nobel Prize in 1956.

2. D. Pines, *Phys. Today* 45 (April 1992): 65.

3. See C. Herring, "Recollections from the Early Years of Solid-State Physics," *Phys. Today* 45 (April 1992): 27. Herring emphasizes that the basic thrusts of Bardeen's initiative are still fundamental to the modern techniques for calculating electron densities introduced by Pierre Hohenberg and Walter Kohn. See P. Hohenberg and W. Kohn, *Phys. Rev.* B136 (1964): 864.

4. Bardeen, *Phys. Rev.* 52 (1937): 688.

5. D. Shoenberg, *Superconductivity* (Cambridge: Cambridge University Press, 1938). This book has appeared in numerous new editions.

6. J. Bardeen, "Zero Point Vibrations and Superconductivity," *Phys. Rev.* 79 (1950): 167.

7. With Philippe Nozières, David Pines is the author of an important work on the N-body problem.

8. Pines, *Phys. Today* 45 (April 1992): 66.

9. T. D. Lee, F. E. Low, and D. Pines, *Phys. Rev.* 90 (1953): 297.

10. Bardeen and Pines, *Phys. Rev.* 99 (1955): 1140.

11. Schrieffer, *Phys. Today* 45 (April 1992): 49.

12. Cooper, *Phys. Rev.* 104 (1956): 1189.

13. Schrieffer, *Phys. Today* 45 (April 1992): 50.

14. Ibid.

15. J. Bardeen, "Development of Concepts in Superconductivity," *Phys. Today* 16 (January 1963): 26.

16. J. Bardeen, L. N. Cooper, and J. R. Schrieffer, *Phys. Rev.* 108 (1957): 1175.

17. V. F. Weisskopf, seminar, December 21, 1979, CERN Publication 79–12 (Geneva: CERN, 1979).

18. After the publication of the BCS theory, the Soviet theorist A. B. Migdal made a remark that would have been devastating had he made it earlier — he showed that any perturbation theory is incompatible with the existence of a gap in the spectrum of electronic excitations (Migdal, *Soviet Phys. JETP* 7 [1958]: 996).

17. THE GOLDEN AGE

1. Jacques Friedel, private conversation, Paris, June 29, 1990.

2. Charles Kittel is the author of an excellent text, *Introduction to Solid State Physics*, 7th ed. (New York: Wiley, 1996).

3. Because neutrons have a magnetic moment, they are deflected by vortex lines in a crystal, which are magnetic. The first studies of the structure of a vortex network came from beautiful neutron diffraction experiments at the Saclay (France) laboratory, where low-energy neutrons were scattered from materials the way light is scattered from an array of slits. See D. Cribier, B. Jacrot, L. Madav Rao, and B. Farnoux, *Phys. Lett.* 9 (1964): 106.

4. I. Giaever, *Phys. Rev. Lett.* 5 (1960): 147.

5. B. D. Josephson, *Phys. Lett.* 1 (1962): 251.

6. The gradient of a quantity such as temperature is a vector, an arrow that points in the direction in which the temperature varies most strongly. The longer the arrow, the bigger the variation.

7. B. S. Deaver and W. M. Fairbank, *Phys. Rev. Lett.* 7 (1961): 43; R. Doll and M. Näbauer, *Phys. Rev. Lett.* 7 (1961): 51. These fundamental experiments confirmed the old prediction of London; only a factor of one-half arising from the Cooper pairs was missing. The methods employed were quite different. Curiously, the two papers were published in the same issue *of Physical Review Letters* as two articles by theorists on the same subject. Lars Onsager, the founding father of vortices in helium, was one of them. The article by John Bardeen, who very much wanted to take part in the event, was published a bit later, in the same volume.

18. AFTER THE GOLDEN AGE: TOMORROW, ALWAYS TOMORROW

1. In an attempt to modernize, the CNRS created a division called "physical sciences for the engineer." This only makes us wonder whether there are other sciences that the engineers can do without.

2. G. Bogner (Siemens-Erlangen), *Proceedings of the European Workshop on High T_c Superconductors and Potential Applications,* July 1–3, 1987, Genoa (Italy), ed. DG12 of the European Economic Community (DGXII/471/87) (Genoa: Fiera del Mare, 1987), 218.

3. A. D. Appleton, in *Proc. Genoa Workshop,* 160.

4. R. Simon and A. Smith, *Superconductors: Conquering Technology's New Frontier* (New York: Plenum Press, 1968), inside cover of dust jacket.

5. Ibid., 154.

6. A large part of the *IBM Journal of Research and Development* for March 1980 (vol. 24) is devoted to different aspects of this program. Of special importance is the article by the leader of the program, J. Matisoo (113).

7. This was not, however, the first attempt to construct computers with superconducting elements. Before the discovery of the Josephson effects, many companies in different countries were making superconducting memories, copies of standard magnetic memories with superconducting elements.

19. THE AGE OF MATERIALS

1. Reported by Anderson in Parks, *Superconductivity,* 2:1351.

2. This *and so on* masks a real difficulty. Even though all physicists understand one another when they speak about first principles, none of them is quite ready to write down a precise list. A cursory poll of some of our distinguished collaborators revealed this, although everyone was happy to include the conservation of energy.

3. Schafroth, his "Australian" competitor, also came from Zurich. Switzerland was well represented in superconductivity, with laboratory groups at the Universities of Geneva and Lausanne and, of course, the IBM laboratory in Zurich, where J. G. Bednorz and K. A. Müller later discovered superconducting ceramics with high critical temperatures.

4. B. T. Matthias, T. H. Geballe, and V. B. Compton, *Rev. Mod. Phys.* 35 (1963): 1.

5. Ibid., 2.

6. J. Labbé, S. Barisic, and J. Friedel, *Phys. Rev. Lett.* 19 (1967): 1039.

7. J. K. Hulm, J. E. Kunzler, and B. T. Matthias, *Phys. Today* 34 (January 1981): 41.

8. M. von Laue, *Ann. Phys. Leipzig* 3 (1948): 31. Bardeen emphasizes this point in his review article on superconductivity in *Handbuch der Physik,* 287.

9. See, for example, R. Comès, M. Lambert, H. Launois, and H. R. Zeller, *Phys. Stat. Sol.* B58 (1973): 587; *Phys. Rev.* B8 (1973): 571; and D. E. Moncton, J. D. Axe, and F. J. Di Salvo, *Phys. Rev. Lett.* 34 (1975): 734.

10. There are several discontinuities depending on the relative orientation of the force and the crystal axes. The phenomenon was first observed by R. O. Bell and G. Rupprecht (*Phys. Rev.* 129 [1963]: 90).

11. R. A. Cowley, W. J. L. Buyers, and G. Dolling, *Solid State Comm.* 7 (1969): 181; J. D. Axe and G. Shirane, *Phys. Today* 26 (September 1973): 32.

12. The stretching was inferred from the doubling of the electron spin resonance line in the presence of paramagnetic impurities. See H. Unoki and T. Sakudo, *J. Phys. Soc. Japan* 23 (1967): 546.

13. Two groups carried out these experiments simultaneously: G. Shirane and Y. Yamada, *Phys. Rev.* 177 (1969): 858; and R. A. Cowley, W. L. Buyers, and G. Dolling, *Solid State Comm.* 7 (1969): 181.

14. R. Chevrel, M. Sergent, and J. Prigent, *J. Solid State Chem.* 3 (1971): 515.

15. B. T. Matthias, M. Marezio, E. Corenzwitt, A. S. Cooper, and H. E. Barz, *Science* 175 (1972): 1465.

16. R. Odernatt, O. Fischer, H. Jones, and G. Bondi, *J. Phys. C. Solid State Phys.* 7 (1974): L13.

17. W. A. Little, *Phys. Rev.* A134 (1964): 1416; V. L. Ginzburg, *Uspeki Fiz. Nauk.* 95 (1968): 91; and 101 (1970): 185. The latter appears in English in *Sov. Phys. Uspeki* 13 (1970): 335.

18. W. A. Little, "Organic Superconductivity: the Duke Connection," in *Near Zero: The New Frontier of Physics*, ed. J. D. Fairbank, B. S. Deaver Jr., C. W. F. Everitt, and P. F. Michelson (New York: W. H. Freeman, 1988), 359. An article on DNA, including a clear image of the DNA molecule, follows this article by Little.

19. *Phys. Today* 24 (August 1971): 23. This reflection of Matthias shows that he would have been quite ill at ease in astrophysics, in which popular conferences have been held these past few years on the so-called fifth force and on magnetic monopoles, objects that have not been proven to exist.

20. A team of Soviet experimenters followed this path: L. I. Buravov, M. L. Khidekel, I. F. Schegolev, and E. B. Yagubshkii (*JETP Lett.* 12 [1970]: 99).

21. W. L. McMillan, *Phys. Rev.* 167 (1968): 331.

22. L. B. Coleman, M. J. Cohen, D. J. Sandman, F. G. Yamagishi, A. F. Garito, and A. J. Heeger, *Solid State Comm.* 12 (1973): 1125.

23. T. E. Philips, T. J. Kistenmacher, J. P. Ferraris, and D. O. Cowan, *Chem. Comm.* 42 (1973): 471; T. J. Kistenmacher, T. E. Philips, and D. O. Cowan, *Acta Cryst.* B30 (1974): 763.

24. The interest in $(SN)_x$, which chemists knew about for dozens of years, arose from a short publication by V. V. Walatka Jr., M. M. Labbes, and J. H. Perlstein in *Phys. Rev. Lett.* 31 (1973): 1139. They presented measurements of resistivity as a function of temperature. They did not reach liquid helium temperatures, no doubt because they were using a commercial apparatus that, operating in a closed circuit like a refrigerator, did not permit temperatures below 10K. An IBM group from San Jose, California — R. L. Greene, G. B. Street, and L. J. Suter — discovered the superconductivity (*Phys. Rev. Lett.* 34 [1975]: 577). Labbes and his collaborators cited the thesis of a chemist from Lyon from whom at least two French physicists were unable to obtain the smallest sample. Since the preparation of $(SN)_x$ could lead to an explosion if it were not done properly, neither of these physicists would take it upon himself to prepare the compound.

25. M. Ribault, G. Benedek, D. Jérôme, and K. Bechgard, *J. Physique Lett.* 41 (1980): L-397.

26. H. R. Ott, H. Rudigier, Z. Fisk, and J. L. Smith, *Phys. Rev. Lett.* 50 (1983): 1595.

27. W. Lieke, U. Rauchschwalbe, C. D. Bredl, and F. Steglich, *J. Appl. Phys.* 53 (1982): 2111; U. Rauchschwalbe, W. Lieke, C. D. Bredl, F. Steglich, J. Aarts, K. M. Martini, and A. C. Mota, *Phys. Rev. Lett.* 49 (1982): 1448.

28. G. R. Stewart, Z. Fisk, J. O. Willis, and J. L. Smith, *Phys. Rev. Lett.* 52 (1984): 679.

29. L. E. De Long, J. G. Huber, K. N. Yang, and M. B. Maple, *Phys. Rev. Lett.* 51 (1983): 312.

20. A SWISS REVOLUTION: THE SUPERCONDUCTING OXIDES

1. This result was announced as late-breaking news in *Phys. Today* 26 (October 1973): 17. The corresponding paper was published only the following year; see L. R. Testardi, J. H. Wernick, and W. A. Royer, *Solid State Comm.* 15 (1974): 1. The temperature achieved was just 0.9K higher than what had just been published for a different experiment; see J. R. Gavaler, *Applied Physics Lett.* 23 (1973): 480. Fifteen years later the publicity machine had taken several steps forward. Then the major large-circulation newspapers would announce on page 1 the new records for critical temperatures. It was, moreover, a sign of indifference that the article of Gavaler was published in an applied physics journal; no one would reasonably think that this result would shake our understanding of how superconductors behave.

2. D. C. Johnston, H. Prakash, and W. H. Zachariasen, *Mat. Res. Bull.* 8 (1973): 777.

3. A. W. Sleight, J. L. Gillson, and P. E. Bierstedt, *Solid State Comm.* 17 (1975): 27.

4. J. D. Axe and G. Shirane, *Phys. Today* 26 (September 1973): 32.

5. Anil Khurana, "Superconductivity Seen above the Boiling Point of Nitrogen," *Phys. Today* 40 (April 1987): 17; the citation of Müller is on p. 18. Anil Khurana was one of the most lucid observers of the excitement that overcame the world of superconductivity at this time.

6. B. K. Chakraverty, *J. Physique Lett.* 40 (1979): L-99; *J. Physique* 42 (1981): 1351.

7. B. K. Chakraverty and C. Schlenker, *J. Physique Colloq.* 37 (1976): C4–353; S. Lakkis, C. Schlenker, B. K. Chakraverty, R. Buder, and M. Marezio, *Phys. Rev.* B14 (1976): 1429; B. K. Chakraverty, M. J. Sienko, and J. Bonnerot, *Phys. Rev.* B17 (1978): 3781.

8. B. K. Chakraverty, *J. Physique Lett.* 40 (1979): L-99.

9. P. B. Allen and R. C. Dynes, *Phys. Rev.* B12 (1975): 905.

10. Charles Wood and David Emin, *Phys. Rev.* B29 (1984): 4582. For an introduction to the subject, see D. Emin, *Phys. Today* 40 (January 1987): 55.

11. B. K. Chakraverty, *J. Physique Lett.* 40 (1979): L-99.

12. B. K. Chakraverty, *J. Physique* 42 (1981): 1351. The original was received by the publisher on August 19, 1979; revised versions were received on January 21, 1980, June 13, 1980, and May 13, 1981. After this long battle the paper was accepted for publication eight days later.

13. J. G. Bednorz, K. A. Müller, H. Arend, and H. Gränicher, *Mat. Res. Bull.* 18 (1980): 181.

14. R. M. Hazen, *Scientific American* 258 (June 1988): 77. The earth's lower mantle is between 670 and 2900 kilometers deep, according to Hazen.

15. Ibid., 74. Perovski was an amateur geologist; he had others do his digging, however, because he was the minister of the interior for the czar.

16. G. Demazeau et al. The article is cited by B. Raveau in *Images de la physique* (Paris: Ed. du CNRS, 1989) and reprinted in *"Nouveaux supraconducteurs,"* ed. H. Pascard, (Paris: Ecole Polytechnique, 1989), 97.

17. Michel and Raveau, *Chim. Min.* 21 (1984): 407.

18. Vera Rich (*Physics World* 7 [July 1994]: 11) reported that, according to *Moscow News*, a weekly in English published in Moscow, three researchers from the Kurchatov Institute of General and Inorganic Chemistry—V. Lazarev, I. Shaplygin, and B. Kakhan—also synthesized a compound of lanthanum and copper oxide. They were hoping to find a substitute for the alloys of precious metals used in electronic circuits for the military. They observed that the electrical resistance diminished with temperature, which brought a skeptical response from their colleagues. Without access to liquid helium, they nevertheless did not dare approach Kapitza's institute with such controversial specimens and gave up their study. If this story were fiction, it wouldn't be credible.

19. J. G. Bednorz and K. A. Müller, "Possible High T_c Superconductivity in the Ba-La-Cu-O System," *Zeit. für Phys.* B64 (1986): 189.

20. M. Tinkham, M. R. Beasley, D. C. Larbalestier, A. F. Clark, and D. K. Finnemore, *Workshop on Problems in Superconductivity* (Copper Mountain, Colo.: Copper Mountain Press, 1983), 13.

21. SUPERCONDUCTIVITY AS THEATER

1. R. Hazen, *The Breakthrough: The Race toward Superconductivity* (New York: Summit Books, 1988), 36. This book is a firsthand account of events from November 1986 to March 18, 1987, from the point of view of Chu's group, as they raced toward high T_c. Hazen describes his own contribution, as he tried to be the first to determine the correct structure of YBaCuO.

2. T. D. Thanh, A. Koma, and S. Tanaka, *Appl. Phys.* 22 (1980): 205.

3. Hazen, *Breakthrough*, 21.

4. Ibid., 22.

5. C. W. Chu, P. H. Hor, R. L. Meng, L. Gao, Z. J. Huang, and Y. Q. Wang, "Evidence for Superconductivity above 40K in the La-Ba-Cu-O Compound System," *Phys. Rev. Lett.* 58 (1987): 405; R. J. Cava, R. B. van Dover, B. Batlogg, and E. A. Rietman, "Bulk Superconductivity at 36K in $La_{1.8}Sr_{0.2}CuO_4$," *Phys. Rev. Lett.* 58 (1987): 408. The latter article was submitted on December 29 and published the following January 26, setting a new record.

6. Hazen, *Breakthrough*, 49.

7. R. Pool, *Science* 241 (1988): 655. This article, written after Hazen's book appeared in the United States, provides a more balanced view of the relative contributions of the Chu and Wu groups.

8. Hazen writes that "during the previous few weeks the superconductor team had obtained evidence that there was an industrial spy working at the Houston department. Documents were seen; phone conversations overheard" (Hazen, *Breakthrough*, 56). He provides no evidence or witnesses. It's also a bit unusual that the spies would call in the results of their espionage from the place where they were spying. On the other hand, it would be customary and not sinister for anyone with an acquaintance in the laboratory to call to find out the latest news.

9. Ibid., 52.

10. Pool, *Science*, 656.

11. Ibid.

12. Hazen, *Breakthrough*, 213.

13. Bogner, *Genoa Workshop*, 218.

22. ALMOST TWENTY YEARS LATER

1. P. W. Anderson, "Superconductivity in High T_c Cuprates: The Cause Is No Longer a Mystery," <http://arXiv.org/abs/cond-mat/201429>.

2. This changes the crystal structure, particularly in the CuO planes.

3. R. Sobolewski, *Supercond. Sci. Tech.* 14 (2002): 994.

4. N. Metzger [*Phys. Today* 55 (January 2002): 54] estimates that private funding of U.S. university physics department research represents less than 8 percent of these departments' resources and is concentrated in a few areas.

5. J. Rowell, M. Beasley, and R. Ralston, *WTEC Report on Electronics Applications of Superconductivity in Japan* (Baltimore: International Research Institute, Loyola College, July 1998).

6. Ibid., 37.

7. Ibid., 44.

8. To be honest, we must admit that European physicists must also provide extensive documentation in order to receive grant funds from European institutions.

9. *Report of the Investigation Committee on the Possibility of Scientific Misconduct in the Work of Hendrik Schön and Coauthors,* <http://www.lucent.com/news_events/researchreview .html>, September 2002. The committee members were M. Beasley (chair), S. Datta, H. Kogelnik, H. Kroemer, and D. Monroe.

10. Ibid., 14.

11. Ibid., 8.

12. See chapter 19.

13. *Science* 295 (2002): 786.

14. Bogner, *Genoa Workshop,* chap. 18, n. 2, p. 548.

FIGURE CREDITS

Every effort was made to obtain permissions, but some rights holders could not be located.

Figures 1.5–6 and 7.2 are from K. Mendelssohn, *The Quest for Absolute Zero*, 2d ed. (London: Taylor and Francis, 1977).

Figures 2.1–3 are from Loek Zuyderduin, Huygens Laboratory, Leiden, the Netherlands.

Figure 3.2 is from Dr. P. M. Claassen, Philips Laboratories, Eindhoven, the Netherlands.

Figure 5.1 is from the Deutsches Museum, Munich, Germany.

Figure 5.2 is from the Kamerlingh Onnes Laboratory, Leiden, the Netherlands.

Figure 6.1 is from the photo collection of Edith London.

Figure 8.1 was reprinted with the permission of Cambridge University Press.

Figures 8.2 and 9.1 are from the Ukrainian Academy of Sciences, Kharkov, Ukraine.

Figure 14.2 is from the North-Holland Publishing Company.

Figure 17.1 appears courtesy of University of Cambridge, Cavendish Laboratory.

Figure 18.1 is from the Société Alsthom-Belfort.

Figure 18.2, top, is from Kaku Karita, Gamma Liaison.

Figure 18.2, bottom, is from David Taylor, Naval Ship R&D Center.

Figure 19.1 is from the National Institute of Standards and Technology.

Figure 21.1 is from the Laboratoire des Solides Irradiés, Commissariat a l'Energie Atomique / Ecole Polytechnique.

Figure 21.2 was reproduced with the permission of AP / Wide World Photos.

INDEX

ABOUT THE AUTHORS

JEAN MATRICON taught physics at the Université Denis Diderot (Paris VII) and did theoretical research on superconductivity and other topics in condensed matter physics. He is now active as a scientific gadfly based in Toulouse and Paris; this is his third book on science for the layperson.

GEORGES WAYSAND has focused his experimental research on superconductivity, most recently at the Université Denis Diderot (Paris VII) and at a spin-off underground physics laboratory he has begun in Rustrel (near Avignon), France. He has been heavily involved in the contemporary debate about the nature of science and the relations between science and society and is the author of several general interest books.

CHARLES GLASHAUSSER is a professor of physics at Rutgers University. His research in nuclear physics is now centered at Jefferson Laboratory in Virginia. He is past chair of the Division of Nuclear Physics of the American Physical Society. He has often enjoyed the hospitality of nuclear physics laboratories near Paris.